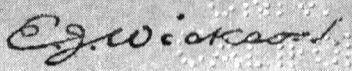

The Suburbanite's Handbook

— OF —

Dwarf Fruit Tree Culture

Their Training and Management

— WITH —

A Discussion on their Adaptability to the Requirements
of the Commercial Orchardist

Both in Connection and in Competition with Standard Trees

By DR. A. W. THORNTON
Ferndale, Whatcom County, Washington
1909

PRESS S. B. IRISH & CO

BELLINGHAM, WASHINGTON

SB355
T45

Gift

Copyright, 1909
By DR. A. W. THORNTON

PREFACE

In blocking out this hand-book I endeavored to place myself in the position of a suburbanite with little or no practical experience on Horticultural subjects, and who was desirous of beautifying and improving his home by the culture of these lovely dwarfs.

Realizing such a one's requirement for a simple, detailed book of instructions, to enable him to know what to do, and how to do it, I present this hand-book for the benefit of suburbanites generally. While many readers may be well informed upon general Horticultural subjects, yet their attention has not been directed to the subject of Dwarf Fruit Tree Culture. I trust they will derive both pleasure and instruction herefrom The work is open to criticism, of course, favorable, or unfavorable, as may happen. Others might have done better, and again they might not. I remember back seventy years ago we had these dwarf trees in our home garden, and, strange to say, that many of the choice fruits of that day still hold a high place in the selected lists of "Bests" in the nurserymen's catalogue of the present day. In spite of the strenuous efforts of three-fourths of a century to surpass them, they still hold their own.

In conclusion I wish to acknowledge the courtesy of Professor Waugh of the Massachusetts Agricultural College in sending some cuts for illustrating this hand-book. Professor Waugh, who is perhaps the best posted man in the United States on the subject of dwarf fruit trees, has written a valuable work on the subject which I can highly recommend

My greatest difficulty in preparing this hand-book occurred when I came to select a list of dwarf fruits, in deciding which to keep in my list, and what to strike out, the claims of many of those stricken out being in many instances fully equal to those retained. Not being able to include all the "Bests" I was compelled to make a selection, and will let it "go at·that," and leave it an open question whether to modify my list in future editions.

In conclusion I will say, I have taken much pleasure in preparing the work, and only hope my readers will enjoy as much pleasure in reading it, and that it may prove the means of attracting their attention to this highly interesting and delightful occupation of Dwarf Fruit Tree Culture.

A. W. THORNTON,
Ferndale, Whatcom County, Washington.

The Suburbanite's Handbook

—OF—

Dwarf Fruit Tree Culture and Management

Bush Pear Tree
Beurre Capaiumont—Photo
Fig. 1.

Apple Tree—8 branches
Trained to goblet form
Fig. 2

It has been found that by treating fruit trees in a particular manner they may be so dwarfed in growth that forty and more apple trees may be grown in the space ordinarily required for a single standard apple tree, at the same time increasing their prolificacy and vastly improving the quality and beauty of the fruit. Other fruits, as apricots, nectarines, pears, plums, etc., are subject to the same change. This dwarfing is no new discovery, but has been practiced successfully in Europe for centuries, and in Japan for a millenium, and has been reduced to a science, that is perfectly simple, and may be successfully practiced by any one who is capable of doing as they are told.

There are three varieties of apple roots which have this power of dwarfing the growth if budded or grafted on them. They are known as the Paradise, Doucin and Crab. The Paradise apple is a slow growing dwarf tree, a native of Europe, and is largely propagated in France, to be used as a stock for working free growing apple scions into, in order to dwarf their growth, and is the best adapted for producing very small trees. The apple trees reduced on this stock are so reduced in size that they may be planted only three or four feet apart, and the bearing age is so forwarded that they will begin to bear some times the first year, and by the fourth year will bear a bushel or more of the choicest quality of fruit. The Doucin apple is another variety of dwarfs wild apple, but is of a more vigorous growth than the Paradise; it is called in England "The broad-leafed Paradise," which causes some confusion in the catalogues of dwarf fruit trees. It is better adapted for apple trees that are to be trained as half standard and espalier tree, as it does not dwarf the growth so much as the French Paradise. Both, however, may be grown in pots, if desired, and yield large crops. The Crab is still more vigorous and is hardly comparable with the Paradise. It is used for growing half standards, and especially adapted for making "fillers" in commercial orchards. All other apples are grown on ordinary apple roots.

The dwarfing of fruit trees is subject to definite laws, which may be briefly expressed thus: "Anything that retards the flow of sap in growing trees has a tendency to dwarf the growth, increase fruitfulness and hasten maturity in bearing." It is therefore evident how peculiarly adapted these dwarf trees are to the requirements of the suburbanite, who on his town lot can have a miniature garden, consisting of forty or fifty of these little trees of the choicest varieties of apples, pears, plums, apricots, nectarines, peaches, figs, grapes and small fruits, not to mention the unalloyed pleasure of tending and training the lovely pets. I do not know of anything more beautiful and interesting than these little trees from the time they first break into bloom in spring and while passing on to the perfecting of their delicious fruit. Above all is the infallible pleasure and pride of the tired and worried business man, or the tired-out society woman going morning or evening to care for the little beauties— a snip here and a pinch there trains them in the way they should go.

Or can you realize the feeling of pride, pleasure and satisfaction after training these little trees with your own hand to grow in pots, and when loaded with gorgeous flowers or luscious fruit, when entertaining your friends, to place pot and tree to decorate your dining table as a center piece, and surprise them with the result of your own handy work This is an experience not uncommon in Europe, where it is frequently practiced. The question of health also is worth considering in this connection Like those little trees producing their fruit so near the ground, secure a degree of health and beauty therefrom not to be obtained otherwise, so the closer the worn-out man or woman can get to work in the ground the happier and better they will feel.

There was a physician in California who was so alive to this fact that he made his female patients believe he could cure them quicker not by giving them medicine, but by prescribing for the vegetables they consumed. He therefore made them grow their own vegetables, fertilizing them with his medicines, which they were to apply to the plants daily, at stated hours, and in strictly regulated quantities; he also succeeded in convincing them that his medicines so altered the character of the juices of the plants that they became entirely different from the stuff they could obtain in the market, and the use of them would quickly effect a cure. When he made his professional calls it was not to see his patients, but to examine how the cabbage, lettuce and cauliflowers were progressing. His patients, of course, got well, as might be expected from the change of lolling in rocking chairs and restricted sunlight to working close down to dear old mother earth, in God's bright sunshine. So with you, the care of these dwarf fruit trees will tone you up more than all the nostrums in the drug store.

To resume, pears are dwarfed by working on quince stock, which enables them to be trained in a variety of forms. Not all pears take kindly to working on the quince, but when they do, they are very satisfactory, and when they do not, we can compel them to do so by the process of double grafting, which is accomplished by first budding or grafting some variety of pear that naturally takes kindly to the quince and then working the rebellious pear on that This has proved a complete success and the result is all that can be desired. The double grafted pears are always of the highest quality (although a little more expensive). Whether owing to the double influence of the combined sap of the quince modified by passing through the pear graft, I know not. The fact remains, however, these double grafted pears are always of the highest quality and well worth the extra price. Root pruning (instructions for which will be given farther on) is also used to check any exuberance of growth. Some times, if too rampant growers, the trees are completely lifted

Photo of Peasgoods Apple and Doyenne du Comice Pear
Fig. 4

and replanted in the same place or removed to another locality, without checking their fruiting. They may also be planted in pots with good effect.

The apricot, nectarine, peach and plum are dwarfed by working on the Myrobolan, Mariana and Mussull plum stocks. They may also be dwarfed to advantage on the "American Western Sand

"Crimson Galande" Peach, 6 years' old
Fig. 37

Cherry." They can be grown in pots, both plain and perforated as well as in baskets (see cuts). When planted in perforated pots, or baskets, the pot is plunged in the spring in a rich border, and the fine protruding fruit fibers feed on the surrounding fertile soil. In the fall, after the fruit has been gathered, the pots are taken up and the protruding roots cut off and the pot and tree removed to the orchard house or cellar. In the case of basket planting, the basket is planted in its place and left there, when it soon decays and leaves the roots free to spread.

The cherry is dwarfed by working on the Mahaleb cherry or the sand cherry. The fig's growth is restricted by potting and root pruning.

Peach in Perforated Pot
Fig. 3

All the above fruits are grown in England and France, and may be grown successfully in the United States if the necessary conditions are complied with. Of course the United States is a mighty BIG country and includes many varying climatic conditions, which may require modifications of treatment for the trees, but there are few regions so inhospitable as to be beyond redemption, as I will presently show. I note the American Pomology Society, in a bulletin (Bulletin No. 8, Division of Pomology), issued by the Department of Agriculture, has divided the United States into 19 Pomological districts, more or less adapted to different varieties of fruits. For the purpose of this hand-book I will reduce that number to the following five:

First—The northern tier of states, consisting of Maine, Vermont, New Hampshire, Wisconsin, Minnesota, North and South Dakota, Montana and Wyoming. This contains some of the most inhospitable fruit regions, but it may to a great extent be made to meet the requirements of those dwarf trees.

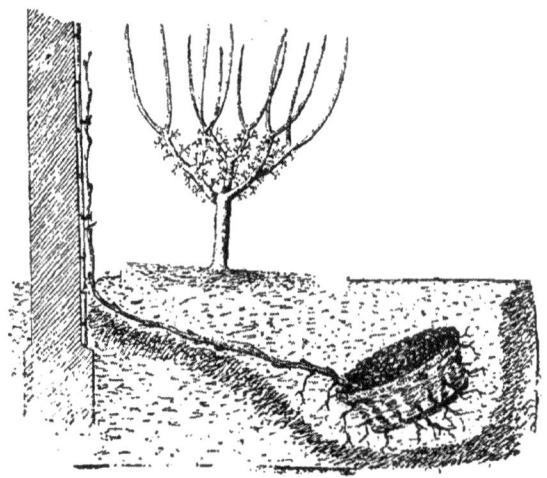

Fancy trained, globlet form
Showing vine when inconvenient to plant near a wall may be planted at a distance in a basket and led underground to wall
Fig. 5

Second—The middle belt of states from the Atlantic coast to the 112th degree of longitude, which comprises a fairly good fruit region, and well adapted to dwarf tree culture.

Third—We have the region of the Rocky Mountains and including Idaho, Eastern Washington and Eastern Oregon and Nevada. This embraces a varied fruit region, in many places producing the finest quality of fruits and in others (from local conditions) some not so good, but nearly all may be utilized for dwarf trees.

Fourth—We have the southern states, with Southern California, which is unadapted to some varieties of dwarf trees while others do well.

Fifth, and Last—We have a region where the dwarf fruit tree garden requirements are met to perfection, namely, Puget Sound, Western Washington and Western Oregon, a region unsurpassed and unsurpassable in many ways, and where every suburbanite should have his dwarf tree garden in full operation to its utmost capacity and enjoy his own apples, apricots, nectarines, peaches, pears, plums, figs, currants, gooseberries, strawberries and grapes.

While thus dividing the United States into fruit sections, no hard and fast rule can be made, and allowance must be made for varying local conditions.

It may be asked here. If dwarf fruit trees are so well adapted to use in the United States and have been grown in Europe for centuries, why have they not been introduced here? They have been frequently introduced and tried, but they were introduced and worked under the European system of management that was not adapted to American conditions. In fact, the introducers tried to open an American lock with an European key that did not fit. Latterly, however, several of the American Agricultural Experiment Stations and some private experimenters have been investigating the subject with good results. To illustrate what apparently trifling errors in details may work injury to the fruit industry. When the practice of training fruit on walls was introduced from England (where it had been successfully practiced for centuries) it was discovered that the trees were quickly killed with the heat. The mystery was not solved for many years, when it was discovered that the difficulty could be obviated by training the trees not against the walls, but to trellises three inches from the wall and thus allowing all of the hot air concentrated by the sun's rays against the wall to escape and secure free ventilation. It is now a fully established fact that dwarf trees can be as successfully produced in the United States as they can in England or France, and the adaptability of these trees to suburbanite's use is freely admitted by experts. The question of their suitability to the requirements of the commercial orchardist is still an unsettled one and open to controversy, with strong arguments in its favor. I will therefore treat the two questions separately, and the reader may judge for himself.

THE DWARF TREE GARDEN FROM THE SUBURBANITE'S POINT OF VIEW.

The suburbanite is generally possessed of a limited piece of ground, and the use of large standard fruit trees is out of the question He requires to combine the ornamental with the useful as far as possible His ground has the great advantage of being well sheltered from harsh winds (a very important consideration in fruit culture). As a rule he does not look for profit from selling his fruit He looks, however, for the enjoyment of beautifying his home and making it attractive to passersby, and if at the same time he can produce fruit of the highest quality for himself, family and friends, he will feel himself amply repaid for the work While he is thus enjoying the pleasures of rural life he is at the same time making a valuable investment by increasing the money value of his property should he at any time desire to sell. Also, if he has children, by giving each of them one or more of these little trees FOR THEIR VERY OWN, and teaching them how to care for them, he may develop a taste for nature studies that will go far to wean them from the streets, hoodlums and other bad influences to which suburbanite boys and girls are exposed While the commercial orchardist requires as few varieties as possible, but in sufficient quantities to furnish carload lots of each fruit, the suburbanite desires as many varieties as possible, though only one or two trees of each kind, so as to secure fresh and varied fruit of his own growing every month in the year. The dwarf fruit tree garden therefore fully meets his wants. He can have a supply of little trees of dessert pears ripening their fruit from July and every succeeding month till the following April; he can also have a few varieties specially adapted for stewing or baking, and can have a few specially suited for exhibition purposes for those dwarf trees will produce the largest and handsomest fruit to be found anywhere. In apples too, he can have a variety of desert apples, ripening every month in the year from July to the following June He can also have a select lot of kitchen apples, lasting from August to the following May, which will add greatly to his enjoyment. In regions that might seem too severe for these fruits, they may be compelled to bear, with a little extra trouble, by growing in pots or boxes, and the luxury of growing them will fully repay any little extra trouble, which, in reality, is no trouble at all, but the most enjoyable kind of pleasure.

Peaches, apricots and nectarines may be classed together. They will ordinarily grow in the open air in many parts of the United States, though in some regions the climate is too severe for them By dwarfing they become hardier, and when grown in pots may be shifted without difficulty, after the fruit has been picked in the fall, to the protection of the orchard house or cellar, and again be set out in the open border to blossom and bear fruit in the summer.

A few plums and cherries should also have a place in the garden, as the can be dwarfed, while the cherries may be saved from the exorbitant toll invariably taken by the birds off high trees. Figs, too, though seldom grown outside of California and the Southern states, can be grown in the open air if given the protection of the cellar in the cold weather and exposed to warm and sheltered spots in the summer Currants, both white, red and black, are very desirable, nor do they take any more room than the dwarf trees. The gooseberry is a fruit not adapted to the hot portions of the United States, but in the cooler regions, is a most luscious fruit to eat out of hand when fully ripe and one that Americans know very little about, judging it from the green, sour, unripe fruit usually seen in our markets. In reality there are varieties of the gooseberry more luscious than any grape when fully ripe. Great attention has been paid to improving this fruit in England both as to size and quality. They now come in a large variety; large, medium and small; red, white, green, yellow; hairy and smooth; late and early in ripening, and if allowed to ripen fully, all are delicious to eat out of hand, and if better known would be more appreciated in this country. For many years great improvement has been made in England in the growth and quality of the gooseberry as well as their training, owing to the practice of giving prizes for the best berries grown each year. This is particularly exemplified in the county of Lancashire, where the vast number of mill operatives are encouraged to compete with one another in producing the finest fruits. One of the greatest improvements is in training as a cordon (see cut) for the trellis or wall. Heretofore it was no joke to pick gooseberries unless one was provided with a good pair of gloves to protect themselves from the thorns, but now by training as cordons, on wires, or walls, this trouble is avoided. So universal is the culture of this fruit in England that leading nursery men furnish lists of over 100 varieties, all having received one or more prizes in different seasons.

Grapes may be grown as dwarfs either in pots or on the California system. There the vines are cut back to mere stubs, each one being shortened annually to only three buds, these buds sending out fruit bearing canes the next season and greatly improving the yield and quality. The grape grown in California is of the "Vitis Vinifera" species, and are of the highest quality, but will only grow on the Pacific Coast in the open air. The American grape is of the "Vitis Labrusce" species, and will not bear the short pruning the California grape stands.

The plum, too, is well adapted to the miniature garden, and is a fruit the best of which is hardly known beyond the Pacific Coast. In Europe and on the Pacific Coast the "Prunus Domestica" is the species chiefly grown; it contains as a class plums of the highest quality, while inferior varieties principally are grown in the Eastern and Western states.

Quinces are a very valuable fruit and well adapted to the miniature fruit garden as it is naturally a slow growing shrub and may be farther dwarfed by root pruning. It is chiefly used as a cooking fruit, making delicious marmalade, jelly and preserves. The small fruits, as strawberries, currants, raspberries and blackberries should all find a place in the suburbanite garden. The strawberry may be grown as a border or edging around the flower beds and vegetable plots, not allowing them to produce any runners, and by planting them a foot or eighteen inches apart in the row they will produce an abundance of fruit I have grown strawberries on this hill plan, keeping the runners clipped off, and have kept the same plants on the same ground for 15 consecutive years and yielding satisfactory crops all the time. This system is not adapted to commercial culture, but fits in to the suburbanite's requirement admirably. Raspberries, especially of the red and yellow varieties, may be controlled and rendered less rampant by pinching the leading bud of each new cane in June. When the canes have reached about three feet in height they will send out side shoots and become more stocky. This pinching of the canes may be continued all through the summer if required to control the growth. I fancy I hear some suburbanite possessing only a small 25-foot lot close in town say: What is this man "giving us?" How can I plant all these fruit trees on my little patch of ground? Wait a bit, my friend. There are suburbanites and suburbanites, some living close in town with their 25-foot lots, and some living further out with lots of one or more acres and all dimensions between who are interested in this subject. Do not repine at your conditions, but take advantage of the opportunities that lie within your reach. If you cannot find room for 50 trees, plant 25; if that is too many, plant 10, or, at all events, try just one, and my word for it, you will be so pleased that you will soon find room for another. Remember, it is the man behind the gun that makes the shooting, be it good or bad. And all these trees are grown under high pressure and artificial conditions. The very aristocracy of the fruit trees (the "400," so to speak) must be treated with all due respect and proper attention. The work is by no means hard or difficult and may be easily accomplished by any intelligent man, woman or child who will obey orders and do as they are told. The requirements must in all cases be done at the right time and in the right manner; it will not do to be satisfied with "I THINK that will do," but go one step farther and say, "There, that is JUST RIGHT."

WHAT TREATMENT DO THESE TREES REQUIRE?

First—They require feeding; the ground must be fertile, and kept so.

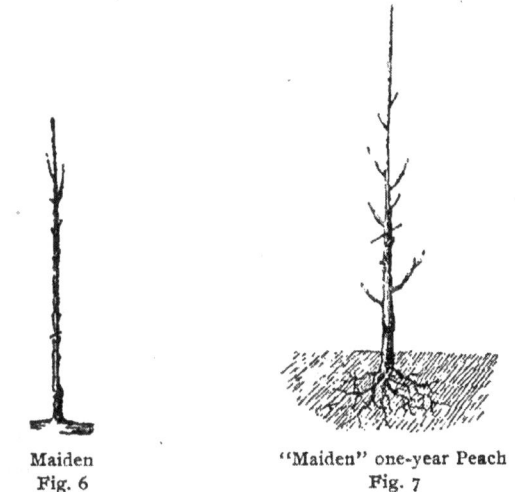

Maiden
Fig. 6

"Maiden" one-year Peach
Fig. 7

Second—They must be kept clean; no weeds must be allowed to rob the land of its fertility and moisture, or the trees will be stinted in their supply of plant food.

OF DWARF FRUIT TREE CULTURE.

Third—They require pruning and training adapted to the system you wish to apply in each particular case. To render this part of my subject more intelligible and comprehensive, I will take each class of fruit separately and discuss the individual requirements of each, even at the expense of some repetition.

First, there are some general questions which require elucidation, such as: Where are these trees to be obtained? The chief source of supply is England and France, where the nursery men keep them in stock at different prices. First "the Maiden;" this is the original dwarfing stock; Paradise, Doucin, Myrobolan, Mahaleb, quince or what not. These are budded or grafted with apple, pear, peach, etc., as desired, but are not pruned in any way and are known as "maidens," or "one year-old trees," and, though small, are the foundation

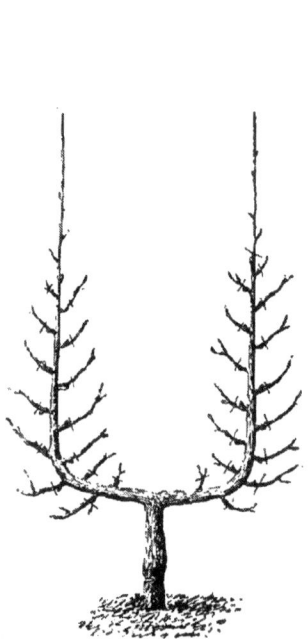

Peach in U Form
2 years' old
Fig. 9

Cordon Apple Trees in bearing, 2 year's old
After Le Cornu
Fig. 8

for all the training of the future tree, as bush, Cordon, Pyramid, Goblet, Palmette, Standard, etc. They consist of a root and stem with the graft or bud inserted, and may be trimmed back to an inch or two of the graft, and are in the best condition for shipping long distances, at the some time the price, duty and cost of importing are at the lowest figure. With these "maidens" the purchaser has entire control of the future form he wishes his little tree to have. Should he not purchase his tree as a "maiden," then the nursery man proceeds to train it to suit himself and sells it the next season as a trained two-year old at an advanced price. It is really a one-year from the bud and may or may not have developed one or more fruit buds. Next season it will develop more fruit spurs, or fruit buds, and the training has been carried on still farther in the required direction, and it is now classed as a bearing tree and sold at a still higher price. After this its cost and value increase year by year in accordance with its size and number of fruit spurs and from training.

The American duty on "nursery stock" is 25 per cent ad valorem, in addition to which it must be remembered there is an "entry fee" charged by the custom house of $2 for each invoice, as well as $1 charge for "permit." It will therefore be seen that it would be inadvisable for the suburbanite to send to Europe for a small lot of these trees. For instance, say he wished to procure $5 worth of trees; he would be required to pay in addition $1.25 duty, $2 entry and $1 per mit; that would be $4.25 in addition to the simple price of the trees, in addition to which there would be freight to New York, insurance and forwarding and overland transportation charges. Nor could he avail himself of the intervention of parcels post service, as the goods would require to be examined at the customs house at the first porty of entry reached in the United States; and however carefully packed originally by the nurseryman, would, under the careless repacking by the customs house people and the further 3,500 miles overland journey by railway, run a very strong risk of being utterly ruined in transit. Some of the stock, however, used for dwarfing if under 3 years old comes at a somewhat lower rate of duty under the class of "seedlings" and "cuttings" which is a specific duty of $1 per 1,000 and 15 per cent ad valorem, together with the $3 entry and permit charges and the trees in this class would require working after they reached here and would be equally unadapted for private importation in small lots.

I may say here by way of parenthesis that as I will be constantly importing these trees from Europe for use in my own nursery to supplant my stock, I shall be pleased to embody in my orders any stock my readers may require at prices for delivery, duty paid, F O B. at Ferndale, to be had upon application

Having now an intelligent idea of what these trees are and where to get them, the next question is: What preparation is required, and this brings us to the consideration of soils, fertilization and planting

Any good fertile garden soil, if well drained, will grow fruit trees; but wet, soggy and lumpy land will prevent success. As I said before, these dwarf trees are high toned aristocrats and require special attention, therefore to secure the best results "intensive culture" is desirable. The land, if possible, should be trenched in the first instance Trenching is performed by first marking the size of your bed, then by digging a trench 18 inches or two feet deep at one end of the bed, taking the soil dug out in a wheelbarrow and dumping it close to but beyond the other end of the bed. You have now a trench from which the soil has been entirely removed to two feet deep; you then continue digging the bed from the trench still two feet deep, turning the first foot of top soil into the bottom of the trench, and the second foot into the same trench on top of the other You now have one trench filled in with top soil at the bottom and another trench open next the undug remainder of the plot. You will continue to dig strip by strip, throwing the soil into the open trenches in front of you, and thus continue until you have dug over the whole plot and have an empty trench left You then throw the soil you dug out of the first trench into this last empty one and you will have your plot all trenched and level. This is the most thorough and best preparation for a garden plot. If you cannot get the whole plot trenched the first year you may take a narrower strip, but wide enough for the trees, and trench it as described and the following year trench an adjoining strip and you will soon have your lot all trenched Should your land not have good soil deep enough to allow you to dig eighteen inches or two feet without striking hard pan, you will require to dig as deep as you can and add a liberal allowance of stable manure, incorporating it well with the soil If your land is reasonably fertile, it will require no fertilizer the first or second year, as you must not force the growth too much. Your object is to check the growth rather than stimulate it, when it reaches the fruiting stage you will then require to stimulate with fertilizers I have been thus particular in detailing the process of trenching because it is to the proper performance of that manipulation that we owe our greatest success in dwarf fruit tree culture. Where the grower has a sufficient area of land that he can avail himself of horse power, he may have the land plowed 12 or 18 inches deep and subsoiled and the surface finely cultivated, as for a garden patch. But above all it must be well drained, either naturally or artificially, as fruit trees will invariably die if they are exposed to cold and wet feet.

We have prepared our ground, and may now get our trees, but what shall we do with them? We must first unpack them, and at this time remember that the roots of these young trees are very susceptible to injury from exposure, so have everything in readiness before you open the package If for any reason you are not ready to plant them permanently, it will be necessary to "heel them in." The expression "heel them in" means to make a temporary planting of them, to secure them from injury until transplanted in their final location, as this is a manipulation that every gardener should be familiar with. I describe it here. Select a spot where no water will stand during the winter, and not having any grass close by to harbor mice, dig a trench deep enough to admit one layer of roots and sloping enough to allow the stems to recline at an angle of about 30 degrees with the ground Having placed one layer of roots in this trench, cover them with MELLOW EARTH EXTENDING WELL UP ON THE BODIES, AND SEE THAT THIS IS FIRMLY PACKED; then add another layer of trees, overlapping the first, and continuing as at first until all are heeled in. As soon as this is done, cover the tops so well with evergreen boughs that they will be thoroughly protected from winds. In sections where the winters are very severe trees procured in the fall can be best cared for in this manner and may be planted out permanently in the spring Having then this trench ready and a pail of water at hand, unpack your trees and look them over If you find them much dried out, dip them in the pail of water and allow them to remain in it a few minutes; if any of the roots are bruised or injured, trim them off with a sharp knife or pruning shears, and "heel them in" as directed, emptying the water over the roots in the trench when about half filled with earth. Be particular at this stage to see the labels are in place and secure from displacement so that the different varieties of trees can be identified in the spring. Be ready to plant the trees as soon as received. We come to the manipulation of planting.

PLANTING—It is better to get your trees in the fall for many reasons, as then the nursery man is not so rushed, and as their supply is not so picked over you can generally secure better trees; and if you have everything in readiness for the permanent planting the trees generally do better, as they start right in to make new fibrous

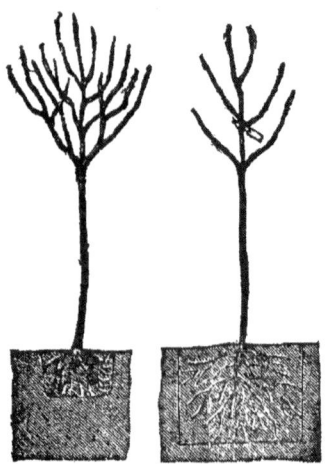

Planting—WRONG.   Planting—RIGHT
Fig. 63   Fig. 64

feeding roots, which, when heeled in, are to some extent injured in transplanting in the spring. Open your package and examine the condition of the roots; if too dry, moisten them; if bruised, trim them; see that the labels are securely attached, but not tied so tightly as to constrict the stems. Before opening the package, however, you must have the holes dug amply large enough to accommodate the roots, spread out in their natural condition, without bending or cramping (see cuts); set the tree in the middle of the hole, keeping it perpendicular; spread out the roots in their natural position, and work in the fine soil with a little stick or your fingers, among the roots until the roots are covered, and tramp them solid (the earth

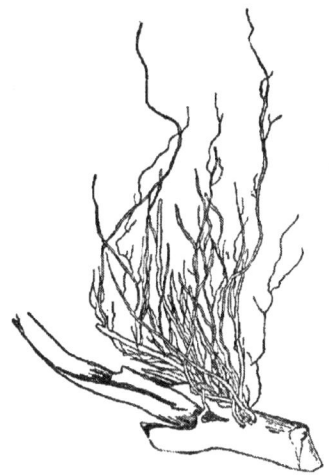
Root development
Root injured before planting
Fig. 10

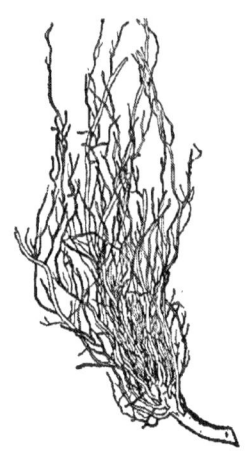
Root development
Not injured before planting
Fig. 11

around the roots must be well compacted), then fill up the hole with loose earth. SET THE TREE FIRM AS A POST, BUT LEAVE THE SURFACE SOIL LIGHT AND LOOSE.

Remember to plant the tree with the point of union between the stock and graft a couple of inches ABOVE the soil. THIS IS IMPORTANT, for although it may be advisable in planting standard orchard trees to place the point of union with the graft below the surface, the dwarf tree require the opposite line of treatment. We graft the free growing cions on the Paradise or other dwarfing stock for the express purpose of restricting its growth, and if we plant the dwarf tree with the point of union below the surface of the ground, the free growing cion will throw out roots of its own and thus antagonize our work for dwarfing. I wish to impress this point on my readers because they will find many authorities recommending the practice of covering the point of union with the soil. This they do from being unfamiliar with the requirements of the dwarf trees, or ignoring their existence altogether; my object in this hand-book being to instruct the suburbanite in the culture of DWARF TREES, not commercial orchards; therefore I say nothing about supporting the tree with a mound of earth or stakes, as at this stage these little trees do not require it. However, we must not forget mulching, which must now be done. Mulching is accomplished by placing a layer of coarse manure, hay, straw or other litter from three to six inches deep, extending one or two feet further all around than the roots. This protects them from the ground drying out, or baking with the wind or sun, and keeps the soil underneath mellow. This mulch may be removed in the spring or turned under and incorporated with the soil in the after culture. Having our trees safely planted, we may now take time to consider the different fruit trees in detail.

APPLES.

As apple trees produce their fruit on fruit spurs, which remain bearing from year to year and for many years, it is of the utmost importance to secure and maintain the largest supply thereof possible and protect them from injury, and next to train the trees into the desired shape. The shape of our dwarf trees will greatly depend upon our special requirements and is in a great measure under our control, though some trees have distinctive habits of growth that may require modification. Thus in bush trees some are naturally close growing and may be planted only four feet apart, while others of a more open habit of growth will require more space and must be planted six feet apart or more. The dwarf apple tree "maiden" is one year old and has not been pruned. When we plant it we cut it back to a point just a little above the point of grafting; the second year it will send out side shoots, and perhaps a few fruit spurs will form the first year from our planting. In June, if there are a number of side shoots, select the most favorably placed for the future frame of the tree and let them grow unchecked till the leaves fall in the winter, when they may be cut back to one-half or two-thirds. The other shoots that start as soon as they have made four good leaves should be pinched back to three perfect leaves, this will have the effect of making a fruit spur in that place and may be expected to bear blossoms and fruit the next season. No shoots must be allowed to grow below the graft as all future growth must be confined to whatever comes from the graft. Sometimes little trees make more fruit spurs than they are able to support, in which case it may be necessary to pinch off some of the fruit blossoms **BEFORE THEY FULLY OPEN**. Thereafter, for bush trees, you may let them grow

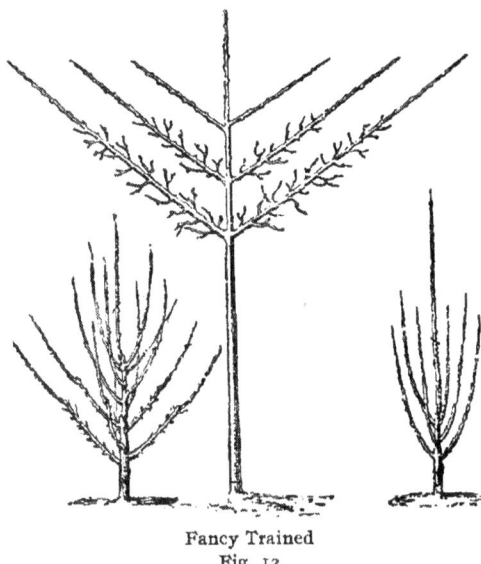

Fancy Trained
Fig. 12

as they will, only pinching back surplus shoots to three leaves to form more fruit spurs. You will bear in mind that all fruit trees that bear their fruit on spurs, when they have their shoots cut back IN SUMMER to half an inch will form fruit spurs, and if pinched back IN SUMMER to three leaves, will do likewise. Every year these little bush trees will bear more and more fruit, the first year producing perhaps one or two, the second year perhaps a dozen, the fourth one bushel, and thereafter increasing crops. It must be remembered also that these fruit are so completely under control that they may be thinned without difficulty to just what the capacity of the tree will justify for production of first quality fruit; they are so dwarfed that the wind has little effect on the fruit in causing windfalls. These dwarfed trees are capable of being trained in a number of different forms, but simple bushes, Espaliers, Pyramids and Cordons are best adapted to the apple (see cuts for different forms of trees). All these forms are the result of training and judicious pruning, and although many of them are the result of pure "fun and fancy," others have very important advantages. It will be observed pruning is of two distinct classes, one for the production of wood and increase of growth, and the other for the restraining growth and the production of fruit, and are known as winter and summer pruning. With dwarf trees the summer pruning is of the greatest importance, a neglect of which will quickly work havoc with your trees. It may be remarked here that there are two distinct systems of pruning. See Figs. Where they are contracted one is generally called the "shortening in" process and may be described as "pruning back from the tips," causing compactness in form, while the other is "pruning out from the stem" and forms a spindling head and is important in stone fruit trees, such as peaches, nectarines and apricots, which produce fruit on last season's shoots.

Instead of growing these trees as simple bushes we can economize space by training them as cordons and at the same time increase their production. Cordons may be either upright oblique or U form

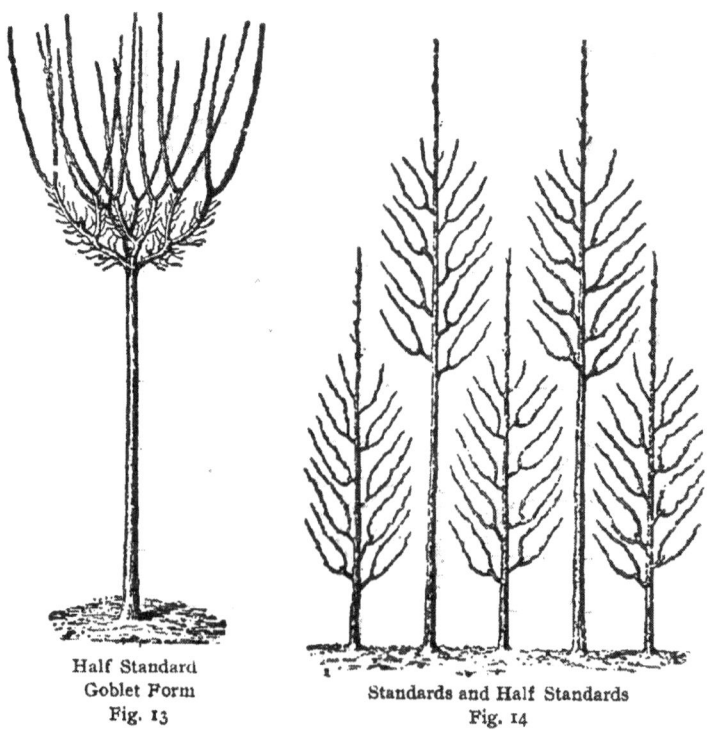

Half Standard
Goblet Form
Fig. 13

Standards and Half Standards
Fig. 14

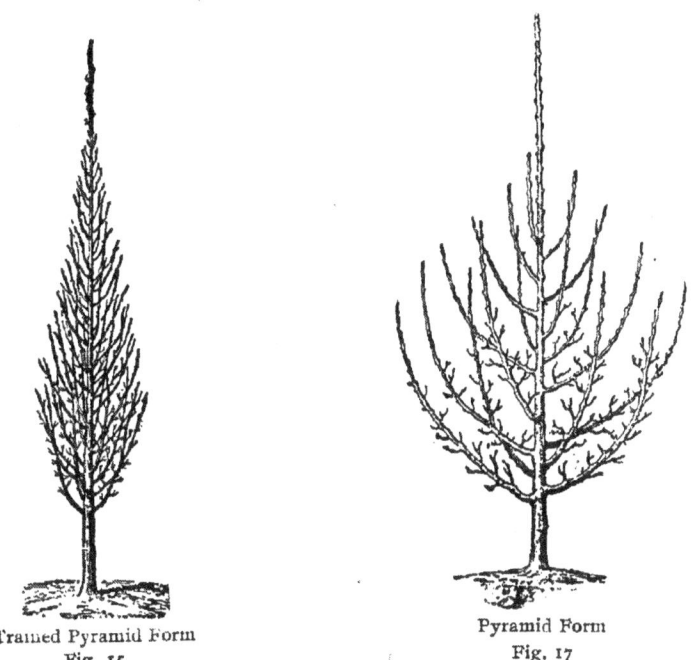

Trained Pyramid Form
Fig. 15

Pyramid Form
Fig. 17

and may be produced as follows: If your maiden tree has been cut back before you receive it, it will require no pruning that winter; in spring shoots will start from the graft buds, of which you will select the most upright growing and tie it to a stake as it grows. In June

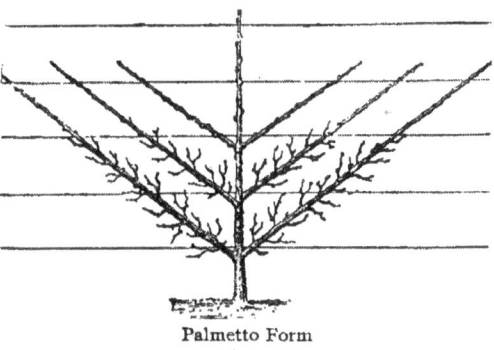

Palmetto Form
Fig. 16

Peach Tree "shortened in"
Properly pruned in summer
Fig. 18

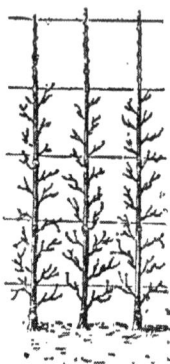

Upright Cordon
2 by 4 ft.
Fig. 20

you will cut all side branches back to one or two inches. In August pinch back any shoots that have made five leaves to three leaves and continue each winter cutting the leader back within one or two

Peach wrongly pruned
Winter pruning
Fig. 19

buds of the last fruit spur and keeping all side shoots cut or pinched back through the summer. Remember the cordon is simply a straight stem without any branches and only leaves and fruit spurs all along its length. If instead of a single cordon you wish to have a double cordon, or U or double U, instead of training one leader perpendicularly, select two opposite shoots and bend them down at right angles and then at six inches farther bend them upright and so continue, leaving six inches between each upright branch, thus a five-branch upright trained tree will measure 24 inches from outside to outside,

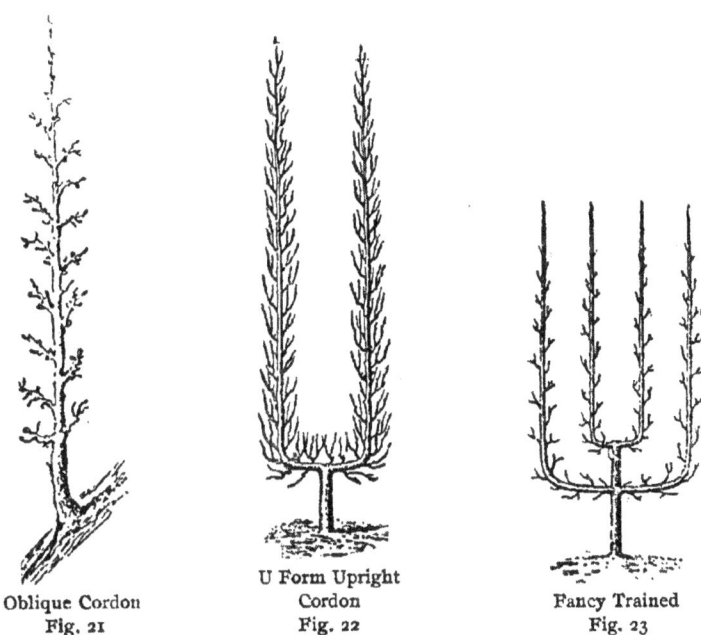

Oblique Cordon
Fig. 21

U Form Upright Cordon
Fig. 22

Fancy Trained
Fig. 23

and consequently may be planted two and one-half to three feet apart. The oblique cordon is between the upright and the horizontal and is intended to secure a longer stretch of bearing wood than would be obtained if trained upright on the same height of wall or trellis. Oblique cordons may be planted only 12 to 18 inches apart. For horizontal cordons plant the trees eight to sixteen feet apart, depending on whether you wish them to be double or single. If single (which is preferable) plant the trees eight feet apart, stretch a stout galvanized iron wire between rigid posts, about one foot

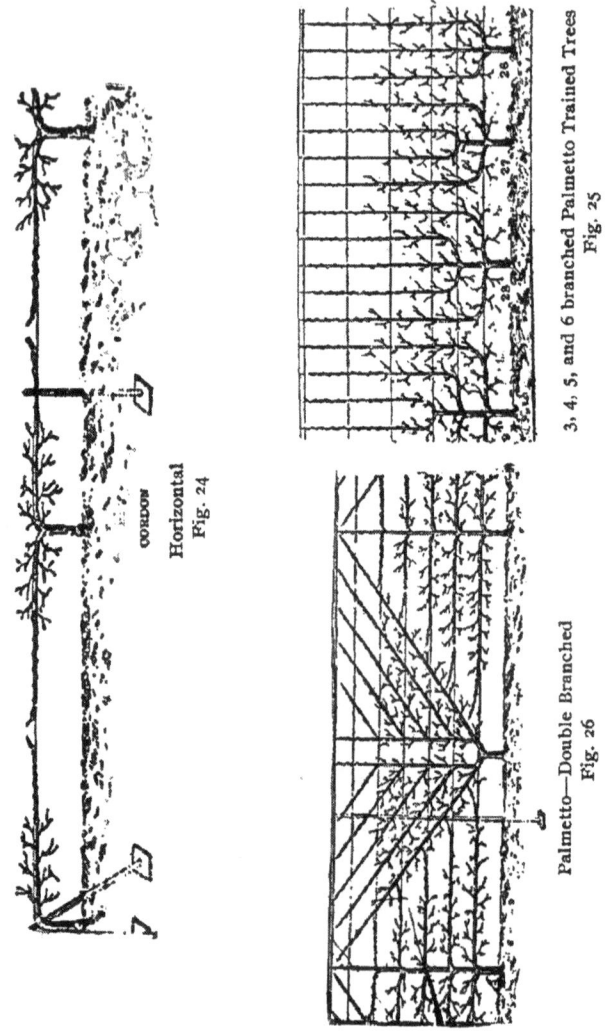

Horizontal Cordon Fig. 24

3, 4, 5, and 6 branched Palmetto Trained Trees Fig. 25

Palmetto—Double Branched Fig. 26

from the ground, and bend the leader at right angles to the stem and train it along the wire as it grows until it overlaps the next tree, when you can graft them together, called enarching, by cutting a little of the bark off each and

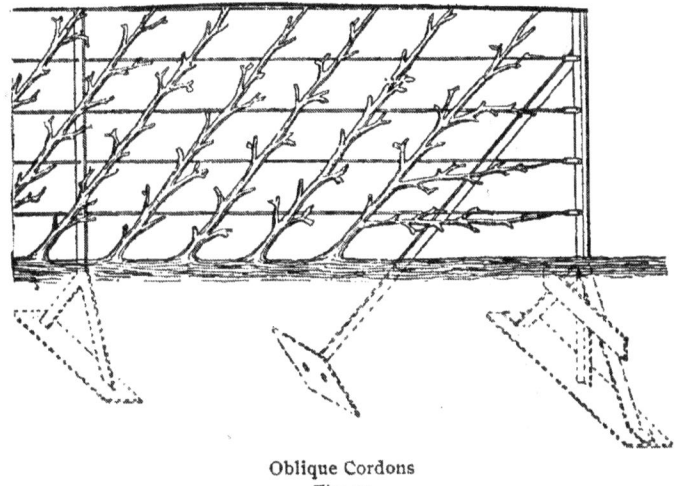

Oblique Cordons
Fig 27

tieing the denuded parts together with grafting wax, making a continuous line of bearing wood all along the wire. This plan is especially adapted for apples and is generally placed along either the front or back of a flower bed. The double horizontal cordon differs only in planting the trees 16 feet apart and training two leaders in opposite directions and tying together the ends of adjoining trees when they overlap. In the latter case you save the cost of one-half the trees required in the first instance. The palmette Verier and the Espallier are merely modifications of the cordon and will be easily understood from the cuts, they have the advantage of supplying a more extended surface of bearing wood and consequently effect a saving in the purchase of trees at the start. The pyramid and goblets are very useful styles of training and are specially adapted to the apple and pear. Pyramids require a very simple system of pruning and yet form the most beautiful and prolific trees for garden or lawn. The whole system consists in simply thinning out the side shoots in June, shorten to half their length in October. In winter a few autumnal shoots will be found to require pruning; these should all be shortened to three or four buds. If the trees are aged or crowded with shoots they should be thinned with a sharp knife; this will constitute the whole pruning for the year. Pyramids should be planted in rows nine feet apart. Goblet or vase-shaped trees are very useful and beautiful; for these a dwarf tree of four or five years old is the foundation if the tree will produce six or eight shoots. For a few years these shoots will require to be tied to stakes for support, but in time will be self supporting. Apple trees of this form are exceedingly ornamental and form beautiful objects either in blossoms or fruit; the hollow center allows the admission of sun and air to the great benefit of the fruit. With these instructions and a fair modification of good taste and attention these varied forms may be produced. The Japanese and some European gardeners produce some very grotesque forms, which have no greater value than the ordinary styles, beyond "fun and fancy," and certainly the forming of them will afford lots of fun.

Half standard apple trees are dwarfed on the Doucin and crab stock, which, while reducing their size considerably, permits a larger growth than the Paradise and render them eminently adapted for use as "fillers" in setting out commercial orchards to occupy space between the larger trees while waiting for them to bear. As the

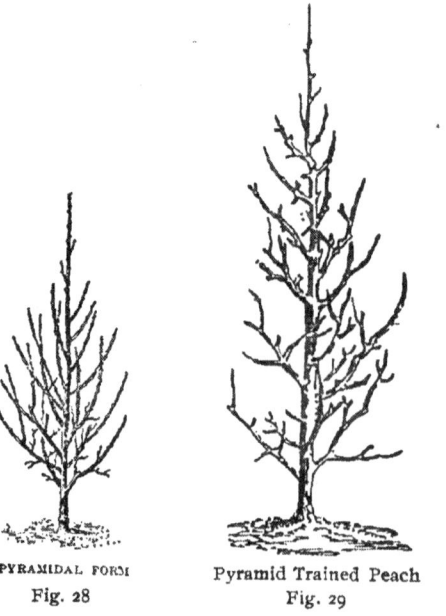

PYRAMIDAL FORM
Fig. 28

Pyramid Trained Peach
Fig. 29

GLOBE FORM
Fig. 29

dwarfs come into bearing so much earlier they will pay a large profit before the large trees begin to bear. Above all things never let any shoots grow on the stem between the root and the graft in any of these trees.

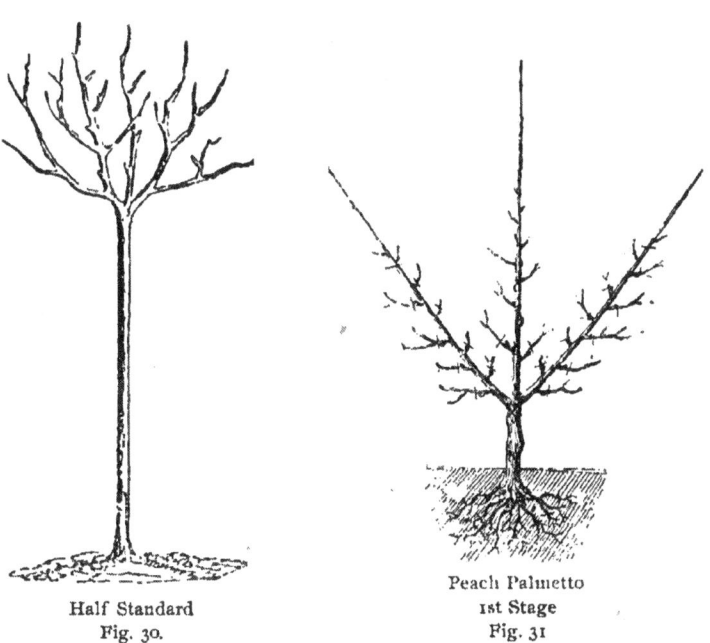

Half Standard
Fig. 30.

Peach Palmetto
1st Stage
Fig. 31

OF DWARF FRUIT TREE CULTURE.   33

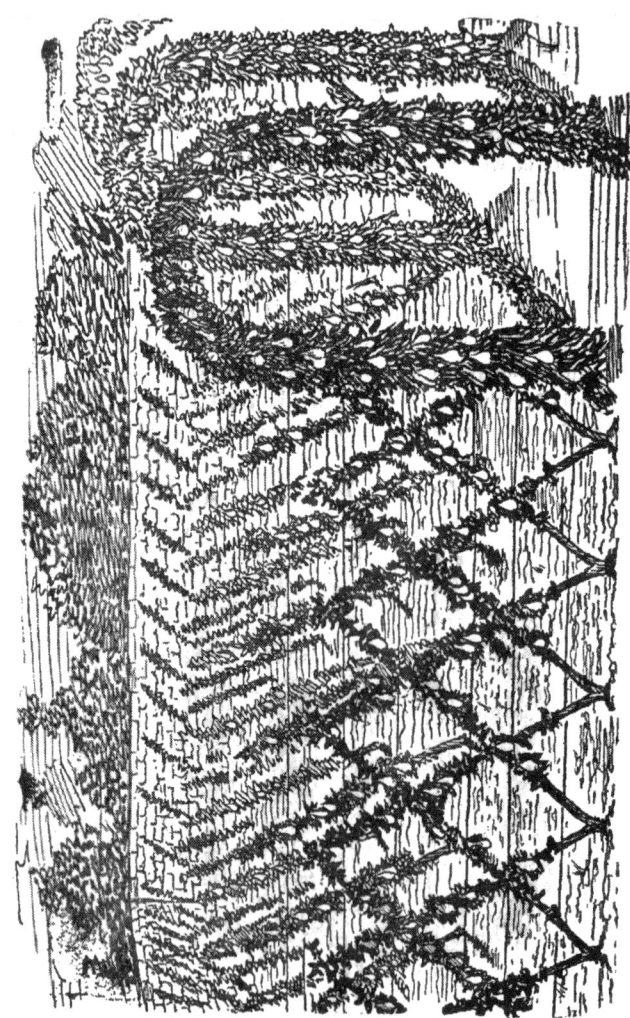

A Profitable Pear Orchard
After Phillip Le Cornu
Fig. 32

PEARS.

The same rule applies to pears as to apples, only pears are dwarfed by working on the quince, by root pruning and pot culture. They may be grown as bushes, pyramids, cordons or half standards. Pyramids and several forms of cordons are best suited to the pear. Walls and trellises also suit this fruit. This fruit is greatly improved by dwarfing and is worthy of all the care bestowed on it. Their season of ripening may be greatly hastened or prolonged in the cooler parts of the United States by winter protection; or be forced by training into cordons and bent back to enable their being covered with a hotbed sash in the spring to protect their blossoms from cold rains and prevent the polen being washed off the flowers. They may be grown as bushes, pyramids, cordons or half standards, but pyramids are the most beautiful and specially adapted to the suburbanite's use, especially where a roomy lawn is available.

As some of my suburbanite readers may like to obtain a money

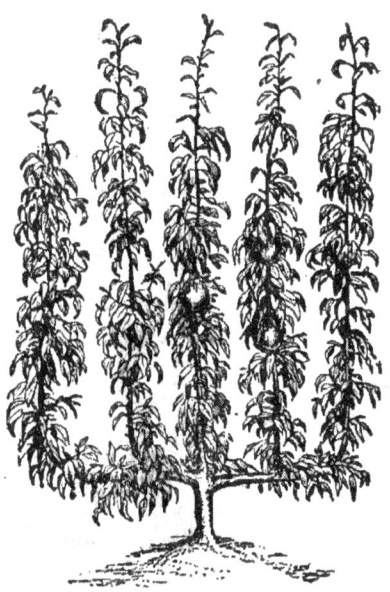

Trained Pear Tree, only 6 inches between branches, 2 ft. from "out to out"
Fig. 33

profit in addition to the pleasures and luxury of having a minature fruit garden, and as both apples and pears are subject to the same treatment, I will in this place give a few hints that may help in that direction. It has become the fashion in England (a fashion that might with advantage be introduced into the United States) for persons with independent means to plant dwarf fruit trees and sell some of the produce to less fortunate neighbors at fancy prices. As an example of the fancy prices that are sometimes paid for dwarf fruit of highly attractive appearance, I may mention that there is a pear grown in Paris called the "Belle Angevine" or "Uvedales St. Germain," so attractive and large that it frequently sell in the high-toned delicatessen stores in the Palais Royale at 30 francs ($5.70) EACH It is of enormous size, often weighing two pounds, and very attractive, but utterly worthless as a desert pear, and as to its cooking qualities I cannot speak, as it is chiefly used to ornament the dinner table, no one thinking of eating it, the ordinary fruit stores selling them for 25 cents each.

On the subject of fertilizers it may not be amiss to quote from Prof. E. Waugh, of the Massachusetts Experiment Station, who has devoted much time and study to this subject, and says:

"While it is true the dwarf fruit trees should be liberally fed, there is a possibility of overdoing it. It has already been explained that the dwarfing of a tree depends in a certain way on its well regulated starvation. If the top could get all the food which its nature calls for, it would not be dwarfed. The rule of feeding dwarf fruit trees therefore should be to give them enough fertilizer to keep them in perfect health and in a good growing condition, but not enough to force unnecessary growth. Fertilizers rich in nitrogen should be especially avoided, and as the object in view is to secure an early maturity of the tree and to produce fruit, always in preference to wood, a larger proportion of potash would naturally be substituted for a diminished proportion of nitrogen. Of course the amounts and proportions of the different elements (nitrogen potash and phosphoric acid) to be applied will vary greatly with different conditions—with the nature of the soil, age of the trees, etc. As a sort of standard we may say that under normal conditions of good soil, with dwarf apple and pear trees in bearing, there should be given annually for each acre:

    400 pounds ground bone.

400 pounds muriate of potash.
100 pounds Perucian guano.

Peaches and plums require more nitrogen during early growth, and more potash when in full bearing. For a new plantation of these trees the following amounts should be given annually for each acre.

300 pounds ground bone.
400 pounds muriate of potash
150 pounds nitrate of soda.

For peach and plum trees in bearing the following formula may be suggested:

400 pounds ground bone.
500 pounds muriate of potash.
100 pounds Purivian guano

Inasmuch as many owners of dwarf fruit trees will have so much less than an acre for treatment, it will be best to repeat these formulae, reducing them to a smaller unit. Making this reduction somewhat freely to avoid long and useless decimals We may compute the quantity needed for each 100 square feet of land as follows:

For Apples and Pears in Bearing:

1 pound ground bone.
1 pound muriate of potash
¼ pound Peruvian guano.

For Peaches and Plums Newly Planted:

¾ pound ground bone.
1 pound muriate of potash
⅜ pound nitrate of soda

For Peaches and Plums in Bearing:

¼ pound Peruvian guano
1¼ pounds muriate of potash.
1 pound ground bone

For treatment of trees in winter, during frost, the trees, if closely packed when received, should be placed in a cellar or some place where the frost cannot reach them, and there remain unopened till a thaw takes place, and then be unpacked and planted; with such treatment, even though frozen solid, they will receive no injury.

If the soil where the trees are to be planted is of fair fertility, no manure need be added before planting, but some fine mold be placed on the roots and the tree shaken so that it enters into the mass of fibers, and then be trodden down firmly. When the hole is filled in level, some manure may be spread on top in a circle of about three feet wide If the soil is poor, some well rotted manure may be worked in when planting. In heavy and wet soils trees should be planted on mounds and not in holes

DISTANCE FOR PLANTING.

Pyramidal pear trees and bushes on quince stock—9 feet apart.
Pyramidal pear trees on pear stock, root pruned—12 feet.
Horizontal Espalier pear trees, on quince, for rails or walls—12 feet.
Upright Espaliers on quince for rails or walls—4 feet
Horizontal Espaliers, on pear stock, for rails or walls—20 feet apart.
Pyramidal plum trees—9 to 12 feet apart.
Espalier plum trees for rails or walls—20 feet apart.
Pyramidal and bush apple trees, on Paradise stock, root prunes, for small gardens—6 feet.
Espalier apple trees, on Paradise stock—12 to 14 feet
The same on crab stock—20 feet.
Peaches and nectarines for walls—15 to 20 feet.
Apricots for walls—20 feet.
Cherries as bushes or pyramids, on Mahaleb stock, root pruned, for small gardens—9 feet apart.
Espalier cherry trees, for rails or walls—15 to 20 feet.
Upright cordons, pear, apple and cherry—2 to 3 feet.
Oblique cordon trees, trained to a wire fence (of four wires five feet high or more) 2 feet apart
Horizontal cordons—single, 5 feet; double, 10 to 16 feet.
Standard currants and gooseberries—6 feet apart.
Cordon gooseberries and currants—9 to 12 inches apart.

These cordon gooseberries, if planted any farther apart, allows too much room for the roots and permits them to grow too rampant and consequently would require root pruning to keep them in bounds If the trees bear too profusely, so as to exhaust themselves, some decomposed manure, about five bushels to 25 square yards, should be spread IN THE WINTER over the surface of the soil and left there. These condensed orchards are for small properties. A small orchard, well cultivated and well planted will be found most productive and profitable. The best form of condensed orchard will be secured by planting oblique cordons, as you will see from the foregoing table of distances for planting that the oblique cordons may be planted only two feet apart in the rows, allowing four to six feet between the rows (which would allow horse culture). We could have 3,630 trees to the acre, and they would commence bearing the second year and bear increasing crops every year after, where there would be room only for 27 standard apple trees at 40 feet apart, as ordinarily recommended in commercial orchards, and moreover the standard trees would not come into profitable bearing for eight or ten years.

Peach Tree in Pot
Fig. 34

Half Standard
Alexandra Noblesse Peach
From Photo
Fig. 35

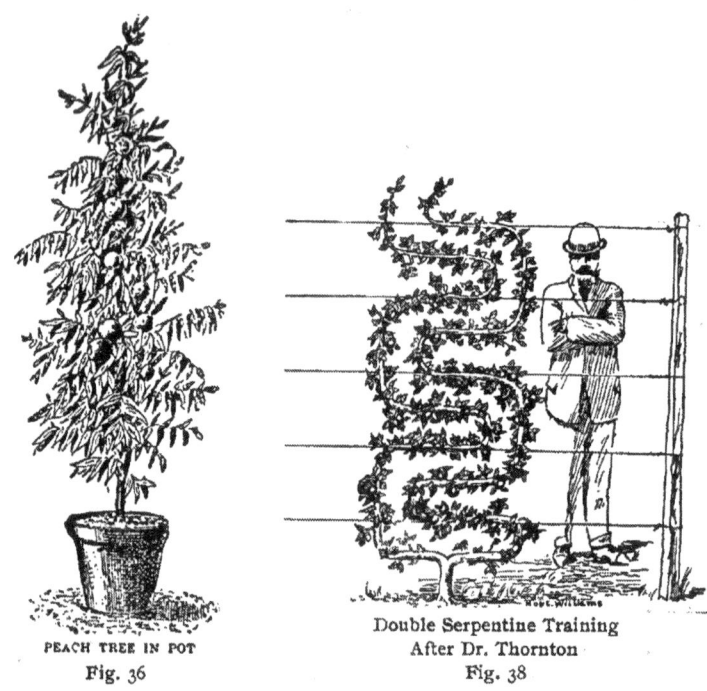

PEACH TREE IN POT
Fig. 36

Double Serpentine Training
After Dr. Thornton
Fig. 38

PEACH, APRICOT AND NECTARINE.

The peach, apricot and nectarine may be considered together, as they require similar treatment. These are more tender than the apple and pear, but are still available for suburbanite's use in most parts of the United States, where the climate is not too severe, as they can stand a considerable degree of frost. They may be grown in pots, either plain or perforated on trellises, or against walls. They produce their fruit on the new shoots, therefore too much of the new growth must not be sacrificed, only enough to let in the light, and control the shape of the tree.

When growing these fruits in pots they will do in 13 to 15-inch pots for the first four or five years, and may be taken up in the spring, repotted in the same pots with fresh soil, and plunged, pot and all, into the border to fruit, or may be transferred from a pot into a rich border and kept there if the temperature keeps above zero. They may be wintered outside with a mulch of straw placed around the roots. If trained to a wall they may be kept close up to the wall in cool climates, but in hot climates it will be better to train them to espaliers or wire trellisses to prevent them getting scorched or burned. Sometimes the foliage becomes too dense, when it will be necessary to clip or pinch off some of the leaves to enable the sun to reach the fruit and brighten the color In this case NEVER PULL the leaves, as doing so will injure the bud adjoining it but pinch with the finger nail or clip with a scissors just below the expansion of the leaf. This may seem a trivial matter, but it is attention to just such trifles that make or mar success with dwarf trees. If potted trees blossom in the house where no bees can get to fertilize them the flower must be hand fertilized; in such cases the blossoms must not be emasculated, as is necessary in hybridizing, which see further on.

PLUMS.

The plum is a very delicious and superior fruit, but is not as well known as other hardy fruits, chiefly because a number of American and Japanese plums of inferior quality have been in use. The plum takes up rather more room in the suburbanite's garden than some other dwarf fruits, as it is generally planted 12 feet apart, as half standards, with a stem four feet high and a round head. It requires little pruning They may however be trained in the same way as other stone fruit, as the peach, and may be confined to more moderate dimensions by root pruning, it bears its fruit on fruit spurs, There are several varieties of plums, prunes, damsons and gages, etc. I would recommend for suburbanite's use the "Prunus Domestica" class as it includes all those of finest quality.

THE FIG.

The fig is a very luscious fruit to eat off the tree, but is very little grown outside of California and the Southern states It is admirably adapted to the small fruit garden, where it can generally be provided with shelter and does not take up much space. It may be grown in pots and shifted into the cellar in the fall. The novelty of growing your own figs adds greatly to the pleasure of doing so

THE CHERRY.

The cherry is admirably adapted to the miniature fruit gardens as they may be dwarfed by working on the Mahaleb cherry or the American Western Sand cherry. They must be root pruned and potted, if required for small gardens, and may be trained in any form and bear fruit on fruit spurs on wood two years old and over. The large Biggareau varieties do especially well on espaliers. Dwarf cherries can also be secured from the depredation of birds more efficiently than on larger trees as they can be covered with netting.

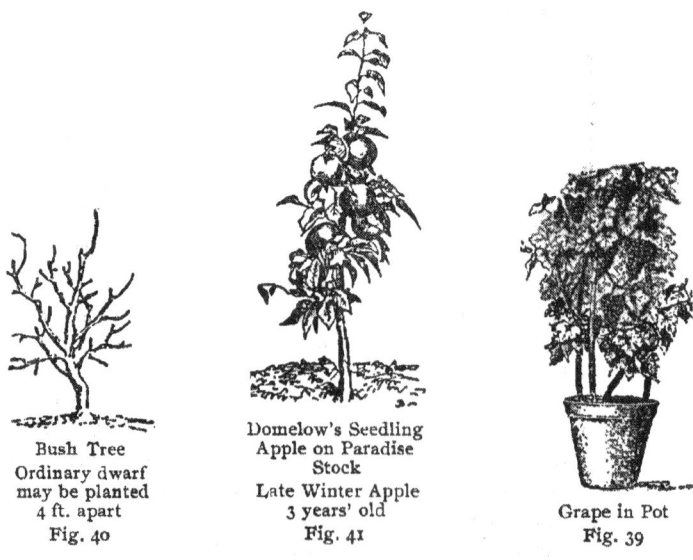

Bush Tree
Ordinary dwarf
may be planted
4 ft. apart
Fig. 40

Domelow's Seedling
Apple on Paradise
Stock
Late Winter Apple
3 years' old
Fig. 41

Grape in Pot
Fig. 39

SMALL FRUITS.

These consist of currants, gooseberries, raspberries, blackberries and strawberries, which no garden should be without. Many of them may be planted between other fruits, or in any out of the way corner. Of currants we have quite a variety—red, white, pink and black—all are good, either for preserving, jelly, wine or to eat fresh with cream and sugar. To produce very large white or red currants the bushes should be closely pruned, the young shoots should be annually shortened to two inches. Currants make very handsome pyramids and bear profusely. Gooseberries furnish a great variety

STANDARD CURRANT
Fig. 42

Currant in Tree Form
Fig. 43

of delicious fruit if allowed to get fully ripe—red, white, green and yellow, smooth and hairy, sweet and acid. For cooking they are generally picked green, but for home cooking it is better to let them get fairly ripe, as they will thus develop a much finer flavor and require much less sugar. With regard to raspberries and blackberries they are so well known as not to require any description here. They bear on new shoots the second year after starting. The canes of the current bear the summer following, and the old canes should be cut away in the fall or winter as they die after fruiting, and only three or four of the strongest new canes allowed to grow. They may be kept in bounds by stopping the new shoots in June.

As to grapes, I will not enter into detail for the reason that so much depends on local conditions that the suburbanite had better consult a local nursery man or fruit grower concerning them. The strawberry has the habit of sending their roots straight down, and do not spread their feeding roots far on either side of the row. This should be remembered when spreading fertilizers. While the commercial grower of strawberries requires to cut down the cultural expenses to the lowest notch, the suburbanite, having only a limited supply of plants, can afford to spend more time and care over them for the sake of producing a higher class of fruit. The strawberry is a fruit that is very sensitive to good treatment, well repays any extra attention given to it. In order to economize space I would recommend the suburbanite to plant in rows as an edging to flower beds, or vegetable plots, putting the plants one foot apart in the rows and keeping all runners religiously cut off; in this way all the strength of the plant goes to the fruit. Should you wish to propagate extra plants of any particular variety it will be better to have a little propagating bed in some out of the way corner; there you can plant the varieties in rows two feet apart and 18 inches between the plants in the row, keeping the ground mellow and free from weeds to encourage the runners to start and root. Another matter about the strawberry is that its size and quality is greatly influenced by local conditions, so that a variety that succeeds well in one place, when grown elsewhere may turn out to be almost worthless, without any fault of the grower or the party who recommends it; it will then be wise for parties thinking of growing strawberries to find out what varieties do best in their immediate locality and not place too much confidence in the roseate descriptions in the fruit calatogs, and only test other varieties in a limited manner. At the same time the converse of this is also true, and strawberries that do not succeed with your neighbor may excel with you. Its attractions are so great we must excuse these little individual peculiarities.

Grape in Pot
Fig. 45.

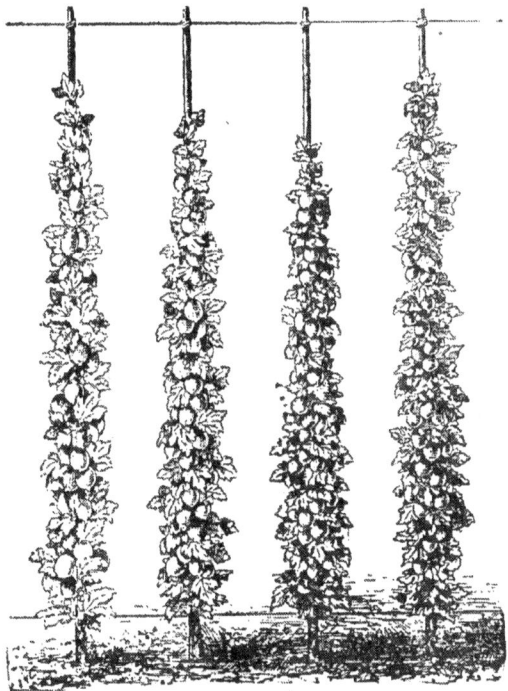

Cordon Gooseberries only 9 inches between the plants
After Le Cornu
Fig. 44

This whole business of caring for a dwarf fruit garden, while affording the highest pleasure and unalloyed enjoyment, has its cares and responsibilities, and although the work is light and easy, it must not be neglected, but must be performed just at the right time, and in the right manner, from the preparation of the ground to the eating of the fruit, for even that most important work must be done just at the right time, or when the fruit is mellow and in its best condition. In June the necessary summer pruning must not be neglected in order to start right; this may mostly be done the first year or two with a pocketknife or the finger nail. A pinch here and a snip there does the work and leads them in the way they should go; but for every pinch and for every snip you should have a definite object in view. First decide what you want, and then stick to that ideal and work to it as near as you can, always keeping in view the necessity of encouraging and preserving the fruit spurs and securing light and ventilation into the heart of the trees.

In former times that was pretty much all there was to be done, but of late years the enemies of fruit trees have so increased that the gardener who wishes to secure the best results must look out.

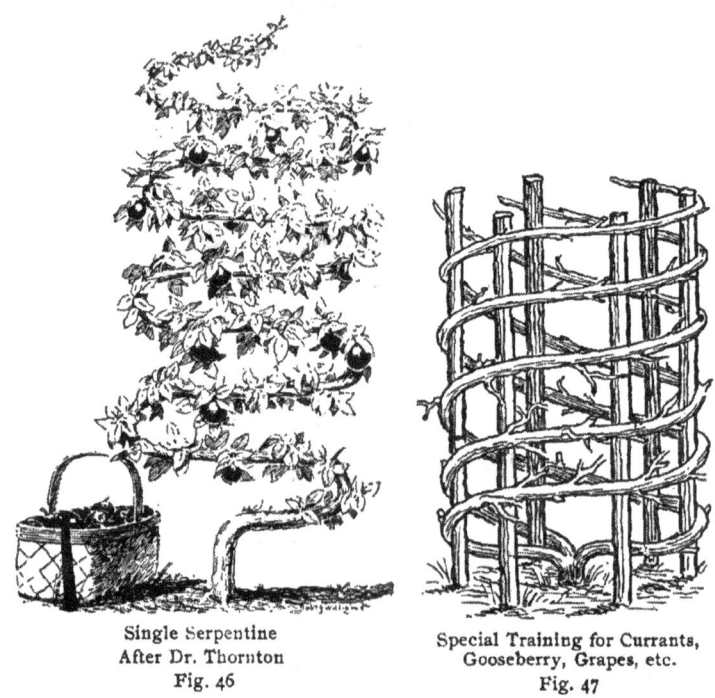

Single Serpentine
After Dr. Thornton
Fig. 46

Special Training for Currants,
Gooseberry, Grapes, etc.
Fig. 47

Fortunately we have efficient means at our disposal, and careful use of them will secure to the grower complete victory over all enemies. Do not let the list frighten you. Among the enemies to the apple are the codling moth, the tent caterpillar and scale; these are all insects enemies and live by eating the fruit and leaves. There are others, such as the woolly aphis and other aphidae, oyster shell bark louse, the San Jose scale that live by suction. This is a very important difference, as one lot requires one kind of poison, while the other requires a different. Thus the biting insects as the codling

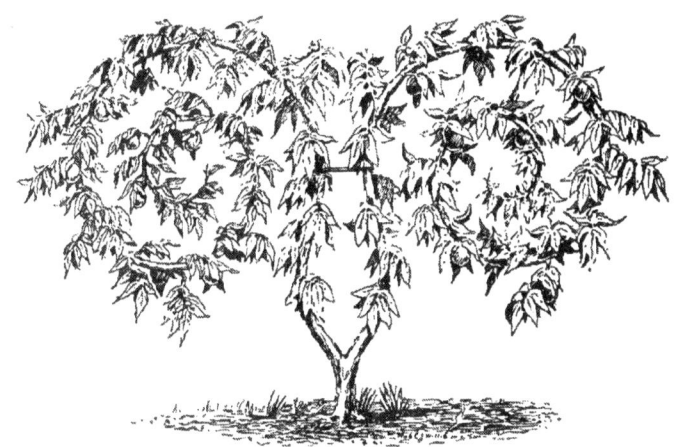

Fancy Trained "Big Horn"
After Thornton
Fig. 48

moth and tent caterpillar require arsenicals, the suckling insects require kerosene emulsion and fumigation. There are germecidal enemies, such as apple scab, brown rot, etc., that require the Bourdeaux mixture or lime and sulphur, or the soluble oils. The pear also has its enemies, as the Phytoptis Pyri, a microscopical insect that causes a blistered leaf; there is also the pear Physilla, both of which may be destroyed by lime and sulphur, or soluble oils. There is also the pear slug, which also affects the cherry. The peach suffers from the curled leaf, and the yellows also rot. The plums suffer from curculie. The gooseberry from milldew, and the currant worm. But enough of this long list of enemies, which all require constant watchfulness and war to the knife. Farther on I will treat at length on this subject.

DWARF TREES IN POTS.

As some of my suburbanite readers may be financially so fixed as to be able to afford themselves the luxury of an orchard house, a few remarks on that subject may be of interest in this place. The orchard house may be an elaborate affair in size, style and finish, or it may be a simple lean-to without heating arrangements and costing only a few dollars. It is desirable to grow the different fruits in different houses, but this is not necessary if one wishes to grow a variety in one house. Only in that case the different varieties should be kept together "en bloc" for the sake of regulating the ventilation to suit each lot. The best form of orchard house is the span roof, not less than 18 feet wide and 4 feet 6 inches to the eaves, and 10

Conference Pear—Photo
Fig. 49

feet to the ridge; in length 20 to 50 feet or more; ventilators 18 inches wide, hinged at the bottom, run along each side of the house, one foot below the eaves; and top ventilators 2 by 3 feet at intervals of 5 feet, alternately, on either side of the ridge. The pots must not be stood directly on the ground, but should have some cinders or broken crockery placed underneath them to insure drainage All fruit trees in pots require to be repotted every year, this may be done as soon as the leaves fall in October. The tree is then taken out of its pot, and the outer soil raked away with a pronged claw till a ball of earth containing the larger roots is left If the tree is healthy and doing well the soil removed will be full of fibrous rootlets. A clean pot of the same size (or one size larger if necessary) having a sufficiency of broken crockery to secure proper drainage, is partly filled with soil to a height that will bring the tree to the same level with the pot rim, as it was before The tree is then placed in the pot, held so that the stem is in the middle, and stands vertical, whilst the soil is rammed firmly in all around the ball and the pot filled up to within an inch of the top of the rim. In potting only a little soil should be used at a time and firmly rammed with a stick before adding more The soil should be of good fibrous turfy loam three-fourths mixed with one-fourth rotten stable manure; for stone fruits, lime in the shape of old mortar, etc., should be added, mix some little time before using and do not allow it to get sodden or dry After they have been repotted the trees should be given some water and stood close together in the house. In severe weather straw should be packed, round and over the pots to keep the frost out Little water need be given the trees in the house during November and December Early in February the trees may be pruned and at the end of the month the trees set four feet apart. A good smoking with tobacco should be given, and the trunks and larger branches brushed with quassia chips wash. If the trees have been properly summer pinched, pruning consists in shortening the last season's growth to behind the point at which it was first pinched. Dead wood and that not required to furnish the tree must be cut out In pruning peach and nectarines the shoots must always be cut to a wood bud (easily distinguished when the flower buds are round and plump and in a triple eye situate between two of these latter) If there are no bees to do it, the flowers must be fertilized or polenized by hand with a soft brush Plenty of air must be allowed at the flowering stage When the fruits are set and the leaves growing the house should be kept closer and the syringe used freely, damping down well at night to obtain a moist growing atmosphere. Peaches and nectarines push too many growths along their shoots; they would be overcrowded if left and must be cut right out, most of the remainder being converted into fruit spurs by pinching out the growing point; only the end bud is allowed to extend, or perhaps one or two others required to cut back to. When stone fruits are beginning to swell they must be cleaned of dead flowers, etc. In most cases the fruits must be thinned out. Pears and apples will, as a rule, thin themselves out, but peaches, nectarines and apricots set too many fruits, all of which would mature if allowed. They must have the crop reduced, going over it three times, once when the fruit has set, again when it is the size of a nut, and finally after stoning is finished.

ESPALIER BEARING FRUIT
Fig. 50

After peaches and nectarines have stoned and when apples and pears are swelling the trees should be top-dressed and given liquid manure diluted with water, about twice a week. A good top-dressing is: Equal parts of horse droppings and loam mixed together, spread out into a bed about a foot deep, and saturate with water. This is ready for use the day after it is made. The mixture is placed on the surface of the soil, about two inches deep near the pot rim and sloping towards the stem of the tree; renew when the fruit is coloring. Summer pinching controls the growth of the trees. When a shoot has made about six inches of growth the tip should be pinched off; the leading shoot of a pyramid may be allowed to extend rather more. The top shoots of a tree (always the most vigorous) are pinched first; this keeps them from taking the lead and keeping it.

Insects must never be allowed to get the upper hand. Aphides are killed by fumigation, directly they are noticed. Red spider thrives in dry heat and is kept down by syringing, forcibly wetting the underside of the leaves and by dampening so as to get a moist atmosphere Syringing must be discontinued when the fruit is approaching maturity. A single four-inch pipe running around the house enables one to keep the frost out when the trees are in flower. In case of a cheap orchard house being decided on, without a regular furnace and piping for artificial heating, the frost can be economically kept out by one of these blue flame kerosene wickless stoves.

Pears in pots form a most useful and satisfactory addition to the fruit garden. The method is simple and certain; an orchard house without artificial heat (either lean-to or span-roof) will shelter the trees until all danger from spring frosts is past. The weather by the end of May, or first part of June, is generally mild enough to enable the grower to put his trees out of doors; after this operation a sufficient supply of water and occasional surface dressing of manure or manure water will insure the production of fine fruit. For the first two or three years of cultivation 13 or 15-inch pots will be large enough; in the autumn, after the fruit is gathered, the trees should be repotted in the same pots and fresh soil added. The trees should then be either replaced in the house for the winter or plunged out of doors, protected against severe frost by covering the surface of the ground with mulch If replaced in the house trees should be also protected during severe frost by a thick covering of straw around and over the pots. At the time of repotting any of the straggling roots should be pruned. When selecting a site for a fruit house for shelter it is advisable for the easy removal of the trees to select a spot with sufficient room to have a summer border in a line with the house. Another method is to grow the trees in perforated pots (see cut).

The border in which they are plunged should be of good soil, mixed with a large proportion of rotted manure, into which the trees will root annually As a rule the rotten manure should be renewed when the trees are repotted, and at the time of plunging the soil should be stirred as deep as the pots are plunged When plunging the pots place a handful of potsherds or gravel under the pot in order that the drainage may be rapid and effectual When taking up these perforated pots in the fall do not forget to cut off the fibrous roots protruding through the pots.

The above system answers equally well for plums and apples. The return is constant and certain, and with plums the fruit bearing season is considerably prolonged as the slight advantage given by the shelter in the spring advances the maturity by fully ten days This result has been repeatedly proven

ROOT PRUNING.

Root pruning is a manipulation little practiced in this country, and very little understood by the horticulturist, but in dwarf tree culture it is frequently absolutely necessary as the only available means of checking a too rampant growth. The important point in dwarf tree culture is keeping the root system completely under control, and changing from the natural wide and deep stretching roots of the ordinary apple tree to a close mass of fibrous feeding roots. Our first effort in effecting this change is the grafting of the free growing cion on some of the natural dwarfing stock, as before mentioned Sometimes, owing to extra fertility of the soil, or other natural cause, the little tree refuses to be controlled and becomes rebellious and starts into a too rampant growth that would upset all our expectations and utterly spoil our work, consequently we are compelled to use heroic measure, which is, in fact, "striking at the root of the trouble." When we first find our little tree obstreperous we give it the first lesson by curtailing its tap root, this is accomplished by forcing a sharp spade obliquely under the roots until the tap root is severed If that lesson is not effectual, we administer the next dose the next season by forcing a sharp spade perpendicularly into the ground at varying distances from the tree, according to its size, and dig in a circle HALF WAY AROUND THE TREE, not turning the soil, but merely cutting the superficial roots. The next year repeat the dose half way around the other side of the tree, Sometimes we dig up the tree entirely, trim the roots and return it back to where it had been growing; this is best done in the fall or winter, and does not interfere with the fruiting the next season. Root pruning is hardly ever necessary in potted trees farther than trimming them if necessary at the annual repotting, as by changing them from smaller to larger pots, as occasion requires; we have the roots entirely under control. The result of all this severe treatment is that our little tree accepts the correction and abandons its evil way and goes to work bearing still more and more beautiful fruit, thus illustrating the wisdom of Solomon in "training the child in the way he should go "

The new course is a very important and interesting phase of dwarf fruit culture, embracing hybridization and cross fertilization, a work particularly adapted to these trees for the reason of their coming into fruit so speedily, consequently by budding the product of our cross fertilization we will be able to produce fruit in two years, and judge of the success and value of our work, while under ordinary conditions we are compelled to wait several years before we can obtain results

Hybridization and cross fertilization consists of removing the polen from the stamen of one flower and placing it on the pistil of another blossom of the same species but different variety, and we take the seeds of the fruit produced by that cross and plant them; the trees that fruit will produce will bear fruit altogether different from either of the original parent's product, and may or may not produce a continuation of the qualities of both They may be better or worse than either or both their parents, larger or smaller, handsomer or less attractive, and at all events a new creation, due to your skill and enterprise Having procured after careful nursing a tree of this hybridized stock, we are naturally anxious to know what it amounts to, but it would take years waiting till that tree naturally bore fruit, which might after all be worthless; or again, a really valuable improvement upon any former product. In the one case you would dig it up and throw it away, or in the other, propogate it to the limit and perhaps make a fortune out of it. All our valuable fruits have been obtained in this way and the originators have been paid large sums of money for the new variety An instance of this come to mind in the case of the celebrated Fay's Prolific Red Currant. Mr. Fay developed this currant by hybridization, and it was so superior to other red currants that Mr. Joslyn, a nursery man, took it in hand and paid Mr Fay some $14,000 in royalties for it I give this from memory and may be subject to some correction. Anyhow, a large amount of money was realized by the originator Now then, we have got a new variety of fruit tree raised and are, of course, desirous to know what it will amount to, so we take a bud from it, when the sap is flowing freely, and insert it into one of the dwarf trees and cause it to develop into a fruit spur, and the next season it will bear new fruit.

The whole process is very simple and interesting and anybody who has a love for flowers can practice the art successfully, for all flowers are subject to the same rule. It was in this way the celebrated Mrs. Lawson carnation was produced, for which Mr. Lawson paid $30,000.

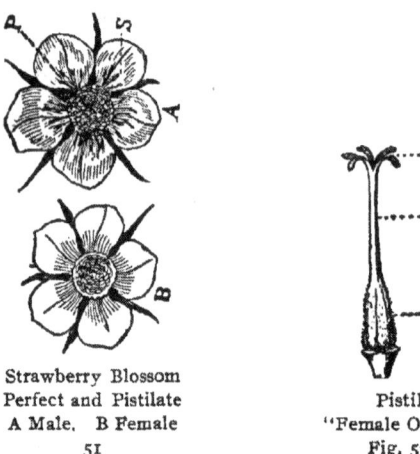

Strawberry Blossom
Perfect and Pistilate
A Male. B Female
51

Pistil
"Female Organ"
Fig. 52

As there are many little details to be observed it may be well to give a cursory sketch of the natural fertilization of flowers. The flower, as every one is aware, is the foundation of the fruit or seed and consist of organs for fertilization. Th stamen is the male organ and produces the polen, which is the fertilizing ingredient. The pistil is the female organ, and at a certain state of its development becomes receptive for the polen, and unless that polen comes in contact with the pistil just at that time there can be no fertile seed produced. Now we have perfect and imperfect flowers, or uni-sexual and bisexual; in some cases, as the strawberry, we have both perfect and imperfect. The perfect plant has both stamens and pistils; the imperfect plant has no stamens, and consequently bears no fruit, unless a staminate plant is growing near it, and the bees and other insects or wind carry the polen from one place to the other. Corn and all nuts have two classes of flowers on the same plant, but differently located. The tassels on the top of the plant of corn are the male flowers, while the silk is the pistil, and at the receptive period the polen is shaken off by the wind and drops on the receptive silk, and the kernel is produced. Now when this polen falls on the receptive pistil it is carried down to the ovary and the pistil then withers away and the seed is developed in due course.

In hybridization and cross fertilizing the following conditions must be observed:

First—The flower must be prevented from fertilizing its own pistil. This is done by clipping off with a fine curved scissors (a manicure or embroidery scissors will answer) all the stamens and corolla, leaving only the pistil standing This is called emasculating the flower.

Second—Means must be taken to prevent fertilization by insects or the wind. This is accomplished by "bagging"—enclosing the pistil after the stamens have been removed in a little bag of tissue paper, closely tied to the branch (or stalk of flower) on which the flower is growing, so that no insects can get inside the paper bags to feerilize the pistil.

Third—When the pistil becomes receptive a slight moisture forms on the top of the pistil (called the stigma), and a watch for that condition must be kept (this generally occurs in the bright warm forenoon), when this is observed a staminate flower from which the polen is to be taken is picked and brought conveniently near the emasculated flower that is to be fertilized. This staminate flower must be in about the same stage of development as the flower to be fertilized. Anyway, it must show the polen in a powdery state on the stamens Then dust the polen on the moist stigma of the pistil either direct from the flower or with a soft camels hair brush; replace the tissue paper capsule for a couple of days and the work is done After the pistil withers there is no further danger of objectionable fertilization, and the tissue paper bag may be removed. It is well to treat several flowers in the same way at the same time to avoid the risk of failure. Next tie a label to the branch that the flower is on, to enable you to identify the fruit later on and make a record of the names of each as to sex This is generally done by naming the female flower first and the male after, thus (Gravenstein and Baldwin). It must be remembered that the result of hybridization will show no difference from the other fruit on the tree, at least not necessarily. It is the seed of that fruit that is altered by the process and the PRODUCT OF THAT SEED will be more or less changed. You must therefore be careful not to permit any one to pick or meddle with the inoculated fruit, or your labor will be lost Most of those hybridized seeds should be planted in moist sand, not kept wet, but not allowed to dry out. Apples, pears, strawberries, mushy small fruit, and hardy stone fruit are generally treated by "stratification," that is, placing them in layers in a box of moist sand, with a cover that will exclude the mice (for mice will find and eat every one of them if they can get at them). They are then placed where the frost WILL GET AT THEM in the winter and may freeze them solid. In the spring they are taken up and planted in a border where they will sprout right away.

I would strongly urge flower-loving ladies to practice the art They will soon become expert, and the enjoyment will be unexcelled They can practice upon their house flowers. The fuscia, for instance, is of a very simple formation and well adapted to practice on; also the tulip and gladiolus, and lily, afterwards they can try some more complicated flowers. The same general principles apply to all. The Salvia Splendens and Salvia Patens have a great promise, the Salvia Patens being the finest blue in the floral world and the Splendens, with its unmatched brilliancy, I believe, have not yet been tested in this way and promise great results.

Nature has many varied and interesting methods of cross fertilizing flowers, some by the action of the wind and gravitation, as in the corn plant, the staminate flower being produced at the top of the plant and the pistilate lower down. When the polen is ripe it falls in a shower on the receptive pistils (the silk). It is also blown about by the wind, so that different varieties of corn planted near each other get "mixed" and the seed will not produce the true type of that originally sown. The cucurbits (or melon, cucumbers, squash, etc.) have the same tendency to "mix" or become cross polenized, they having the staminate and pistilate blossoms on the same plant, but separate from another, and in this case the staminate flowers vastly outnumber the pistilate On the other hand, the holly has the two classes of flowers on different trees, and the tree will not bear its beautiful scarlet berries unless it has perfect flowers or has a staminate flowered tree in its vicinity. Some plants (such as the sweet pea and others of the same family) fertilize themselves before the flower opens, and consequently do not get "mixed," if growing close together. This is important to the hybridizer, as showing the necessity of emasculating the flower to be hybridized before they fully open to prevent self fertilization. Again, some fruit have not the power of self fertilization, as the Bartlett and Beurre d'Anjou pear. A remarkable illustration of this peculiarity occurred some years ago in Oregon, near Salem, where a gentleman came into possession of 160 acres and began to cast about what to do with it. At that time Oregon apples and pears had a high reputation on the Pacific Coast, and he interviewed some of his neighbors who had commercial orchards, and they advised him NOT to have many varieties, but have enough of one variety to produce carload lots of fruit of a kind. This was good, sound advice, as far as it went. Our friend studied over the matter and found that most of the orchards were chiefly given over to apple culture. He also found that not very many pears were grown in his vicinity, and that pears fetched a higher price in the market than apples, and that the Bartlett pear always stood at the head of the market. As he had plenty of money to enable him to indulge his own desires, he decided to plant his 160 acres with a solid block of Bartlett pears. Consequently he gave a contract to a local nursery man to furnish and plant 5,500 Bartlett pears, which was accomplished in first-class style. It was a picture to see those trees growing in rows half a mile long and as straight as a line could make them. The gentleman took the greatest pride in his pear orchard, keeping it well cultivated, not allowing a weed to grow on the whole quarter section, and waited for the time for fruitage to come. But alas, no fruit came, and, unlike the House of Israel, described by the prophet as a vineyard that brought forth wild grapes, this pear orchard did not even bring forth wild pears, but was utterly barren and unproductive. He now thought it well to see what the Agricultural college men had to say about it, which he should have done before he started in. The first question asked by the professor was: "Were there only Bartlett pears in the block?" "Yes, only Bartlett pears, and they were always so thrifty." "There was your error. The Bartlett pear is not self fertile and requires other varieties planted in close proximity to fertilize the flowers." Consequently he had to dig up or graft over a large number of the trees and plant other varieties.

The fig is an example of a very interesting peculiarity in fertilization. There are three classes of fig trees, the Capri (or wild fig), growing in Symra; the fruit bearing fig, growing also in Smyrna, and the Adriatic. There is a peculiar kind of wasp that breeds in the Capri or wild fig, and unless those Capril figs at the proper season are removed and hung up in the Smyrna fig tree it will not be fertilized. The Adriatic fig, not having any Capril figs near, cannot be fertilized and only produce an inferior class of figs, but its seed is non-productive. The flowers of the fig are inside the fruit, and to be fertilized the wasp crawls in and fertilizes them; consequently the best figs in the market are (or rather have been) the Smyrna fig. In California in the early days the mission padres imported the inferior Adriatic figs. Of late years, however, the Capri fig and its little wasp has been introduced into that state, and now the production of the true Smyrna fig is an accomplished fact in California. The California Smyrna fig outranks the imported variety.

GRAFTING AND BUDDING.

The necessity of testing the result of our hybridizing as soon as possible leads up to the subject of grafting and budding, which every horticulturist should understand and be able to practice. I will now describe the "modus operandi" thereof.

Cleft Grafting
Fig. 53

Cion for Grafting
Fig. 54

Grafting and budding are modifications of the same process and have the same object in view; that is, to reproduce a variety of fruit or flower from the bud of one already in existence. Grafting is of several varieties, such as root grafting, crown grafting, whip grafting, wedge or cleft grafting and shield grafting. For suburbanite's use, however, the cleft and whip grafting are the only varieties likely to be practiced. Root grafting is done in the winter, and may be done in the house and by the fireside and packed in a box of moist earth and kept in the cellar or buried till the spring and then planted in the nursery. In the case of the apple, pieces of apple roots, about four to six inches long and about the thickness of a lead pencil, are secured as a stock, and cions of the same thickness are grafted in the same manner as whip grafting, which will be described further on. It is chiefly used by nurserymen for propogating large quantities of nursery stock during the dormant season, when other work is not so pressing, and several pieces of root may be taken from the same tree. While crown grafting requires a whole root for one graft and is intended to be set below the ground where the cion will throw out roots of its own in addition to the original root it is grafted into. The cleft grafting is done in the spring, when the sap begins to flow in the stock, the cions having been cut in the dormant season and stuck in a box of moist soil in the cellar, or are simply stuck in the ground at pruning time, preferably in the shelter of a north wall, the object being to keep them back from sprouting till after the stock has fairly started. The stock is cut off with a sharp pruning knife or shears, and taking a sharp chisel and malet the stock is split sufficiently to allow the

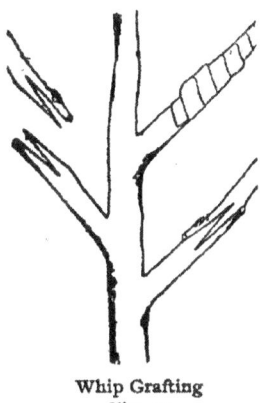

Whip Grafting
Fig. 55

cion, after it has been cut sloping or wedge shaped, to be inserted into the cleft in the stock, taking care that the cut edges of the shaved cion are accurately adjusted to the edge of the bark in the stock. Generally, the split stock will hold the cion sufficiently secure without tying. The whole point of union is filled up and covered with grafting wax to protect the recently cut portions from exposure to wind, water or sunshine (see cuts illustrating this subject). Sometimes it is convenient to bind up the joint with waxed string, described further on. Whip grafting is used for small branches or young seedlings, where the cion should be of about the same diameter as the stock. Both are cut slanting, and with a harp knife a split is cut downward in the stock, commencing at the middle point of the sloping cut; a similar split is made upwards in the cion, and both are locked in each other, as shown in cut, taking care to have the inner line of bark of each in accurate adjustment at least on one side. The whole length of the joint is bound tightly with waxed string, the cion, with only one sound bud, projecting above the waxed wrapping. In the cleft grafting it is well to put in two cions, one on each side, and after they have fairly started to grow, one shoot can be cut off, leaving only one to obtain all the sap and strengthen its growth.

The great advantage of grafting as compared with budding consists in the fact that if from any cause the graft fails to grow, being done in the early spring, one can bud the same stock in the summer and save the loss of a whole season. The stone fruits are generally better adapted to budding. Both grafting and budding are equally adapted to floral work as to fruits.

The operation of budding, although for the same purpose as grafting, is distinctly different in the modus operandi. In the first place, it is performed in summer when the sap is in full flow and when the bark will lift freely from the wood, both of the stock and bud stick. The process is simple, however, and is specially useful in training dwarf trees to secure uniformity and a balance of growth. For instance, in training Palmetto forms and other fancy shapes we are often hindered by side shoots not starting where we desire them, insomuch that the nursery man or private cultivator will start to make his tree of a particular form, when the willful little thing takes the notion to follow its own sweet will, regardless of consequences, and it becomes less trouble to give way and let it take its own natural form than to fight it out; nevertheless by budding we can enforce the growth of a shoot just where we desire it to be. This will be apparent from the cuts. Sometimes in growing cordons the stems fail to furnish sufficiently with fruit spurs and we can then put in one or more buds in any position along its stem.

Having selected a suitable stock for budding, it is necessary to procure sufficient plump buds; these are generally taken of the new or current season's growth by cutting off a shoot with several buds upon it; this is called a "bud stick." Of course only one bud is used in a place, but as many may be inserted as we find suitable room for We now select a smooth spot in the bark of young wood and cut a T down to the cambrium or sap wood, lift the bark on both sides of the upright cut, then we take the "bud stick" and selecting a plump bud, make a horizontal incision about one-eighth of an inch above the top of the bud, then turning the bud stick with its top pointing downwards, make a cut beginning one-fourth to one-half inch below the base of the bud and cut upwards and deeper than the bud, until the cut meets the horizontal cut already made and a little shield is separated with a square top and the bud in the center. Sometimes we will find a little spicula of wood still adherent to the back of the bud with a little "nipple" of soft wood entering the base of the bud. By inserting the point of a knife under the lower end of this wood it is easily lifted free from the bark; some times it will stick pretty firmly. It is a mooted question whether it is best to remove it; some do, which is the English system, and again some do not, which is the American As far as my experience teaches I do not think it matters very materially If I can remove it without injury to the bark of the bud, and it lifts easily, I generally take it away; if not, I leave it in its natural position and I find no great difference in the result. We now have a little shield of bark, with an uninjured bud thereon, and we lift up the flaps of loosened bark on the stock and slip the bark with the bud on under the loosened bark, being careful to adjust the upper end of the bark shield accurately to the horizontal cut on the stock. THIS IS IMPORTANT. Then tie the bark tightly, both above and below the bud, with a soft string, no grafting wax being necessary. It is well if possible to have the stem of the leaf attached to the bud, only clipping off the expanded part, as it will assist in handling the bud, and by its condition in a few days will show whether the bud has "taken" or not. We now leave the bud alone for a week or ten days, when the string must be loosened or entirely removed.

"The reason why" will give an intelligent idea of the process. In grafting the work is done in the spring and the graft grows and is nourished by the ASCENDING SAP. While in budding the work being done in summer, the new bud remains dormant, but is nourished by the DESCENDING SAP that is elaborated in the leaves as it flows downward to nourish the roots. The sap flowing between the "cambrium" (or sap wood) and the inside of the bark shows the necessity of having the inner bark of both stock and cion accurately adjusted to facilitate the flow of sap from one to the other. A careful study of the accompanying cuts will better eluci-

date this branch of my subject than pages of letter press, and here
I would recommend to my readers to take advantage of any opportunity they may have to get someone knowing how to show them
the manipulation of the process One practical example will be
ample instruction, and success will be the result of practice and
painstaking.

GRAFTING WAX is made after many formula, but one of the
best is· Take of tallow 1 oz., bees' wax 2 oz., rosin 4 oz., melt all
together into a uniform fluid condition by stirring and pour into
cold water; when cool enough to handle, having first greased your
hands, pull it as if pulling candy until it attains a straw color and
roll into convenient stick, when it will harden, and keep in a cool
place. When requiring to use it, again grease your hands and work
it up until sufficiently softened, and press it with your fingers
close around and filling all crevices about the point of junction of
stock and cion, from which it need not be removed as it will gradually wear away as the tree grows

GRAFTING CLOTH is more convenient in many ways and is
made by tearing strips about one inch wide of any old materials at
hand, similar to rags used for making a rag carpet These strips,
however, need not be sown together, but rolled as tape or a bandage
is rolled, the end of one piece simply overlapping the other. When
the roll is sufficient size for handling, one and one-half to two
inches in diameter, the free end of the last strip is tied loosely on the
roll with a thread to prevent unrolling, and is thrown into a vessel
of hot melted grafting wax, where, by stirring it round and squeezing
it with a stick, it will become saturated with the hot wax and may
be taken after squeezing out excess of wax and laid aside to cool.
When required for use sufficient of the strip of waxed cloth is unrolled and wound around the graft·in a spiral manner, each turn
overlapping the previous one-half or one-quarter inch, when the
union will take place underneath As the branch or graft grows.
if it shows any sign of contraction or swelling above or below the
wrapping it must be slackened sufficiently to prevent strangulation
The importance to the suburbanite of a knowledge of budding and
grafting will be seen from an examination of the various forms of
training dwarf trees shown in the cuts Much of the beauty of
those trained trees depends upon the success gained in securing
uniformity and balance in the product, especially in the Palmetto
and Vernier forms. If we fail to secure shoots for frame of tree

exactly where desired, the tree will become lop-sided and ungraceful, and as sometimes these trees fail to put out shoots exactly in balance we can secure uniformity by inserting one or more buds just where required, and thus "save its face," which, without the knowledge of grafting and budding, we would be compelled to relegate the specimen to the less attractive form of the ordinary bush in a "happy go lucky" or "hit or miss" style. Again, in case of failure from any cause of one of these trees to furnish fruit spurs along the stem and leaving irregular bare spots unfurnished, we can amend the fault by budding where required.

Bush tree 4 years' old
"Cox's Orange Pippin
Fig. 56

Pyrimid in October
From Photo
Fig. 57

DWARF FRUIT TREES FROM A COMMERCIAL FRUIT GROWER'S POINT OF VIEW.

While the adaptability of dwarf fruit trees to the suburbanite's requirements is now an established fact, its applicability to the conditions of the commercial orchardist is still open to controversy and worthy of consideration. While the suburbanite may have little or no experience, the commercial orchardist knows just what he wants and can form as sound an opinion for himself and is just as capable of weighing arguments that I submit as I am For a numbers of years I have been a commercial orchardist and have fought the fruit pests in every available manner. I remember long ago, before fruit pests had become so multiplied and spraying was invented, that we grew fine fruit with little difficulty. We had, of course, the codling moth and the curculio and a host of other fruit pests, but nothing to compare with the present condition of things, and there seems to be little show of improvement with all our advanced knowledge and extra work. What we want is to be able to reduce the amount of work and make what is absolutely necessary easier. We require to reduce the size of our trees to reduce the labor of thinning, spraying, picking, lessen the number of windfalls, increase the yield of fruit, and, above all, improve the quality and beauty of our fruit, and early bearing of the trees, and thereby increase the prices and profits of our orchards. All these objects may be attained by the intelligent adoption into our system of orchard management of the use of dwarf fruit trees.

It has been demonstrated beyond question that the dwarfing of fruit trees has the effect of increasing the prolificacy and early bearing as well as the size, beauty and quality of the fruit. It has been found that these apple trees dwarfed on Paradise stock will begin to bear the second year from the bud, sometimes even the first year, and by the fourth, will frequently bear one bushel or more of choice apples It must be remembered that these little trees may be planted only four feet apart, and under some circumstances even less; they may be taken up and moved from place to place, and from time to time, without interrupting their fruit bearing. The following is a report of an experiment in growing apples dwarfed on Paradise stock to establish their yield in England:

Apple tree planted (a Warner's King).
1871—Planted a "maiden" tree.

1872—Bore three apples, first year.
1873—Bore 1½ peck, second year.
1874—Bore 2 peck, third year.
1875—Bore 4 peck (1 bushel), fourth year.
1876—Bore 6 peck (1½ bushels), fifth year.
1877—Bore 7 peck (1¾ bushels), sixth year.
Total in first six years 20½ peck (5 bushels).

Now in view of these figures let us make a comparison between one standard apple tree and a block of Paradise dwarfs, occupying the same space. We will say a standard apple tree (to do its best work) should be planted 40 feet apart, that is requiring 1,600 square feet or 27 trees to the acre, while the Paradise trees are 2, 3, 4, 6 or 9 feet apart, according to the style adopted. Let us take as a basis

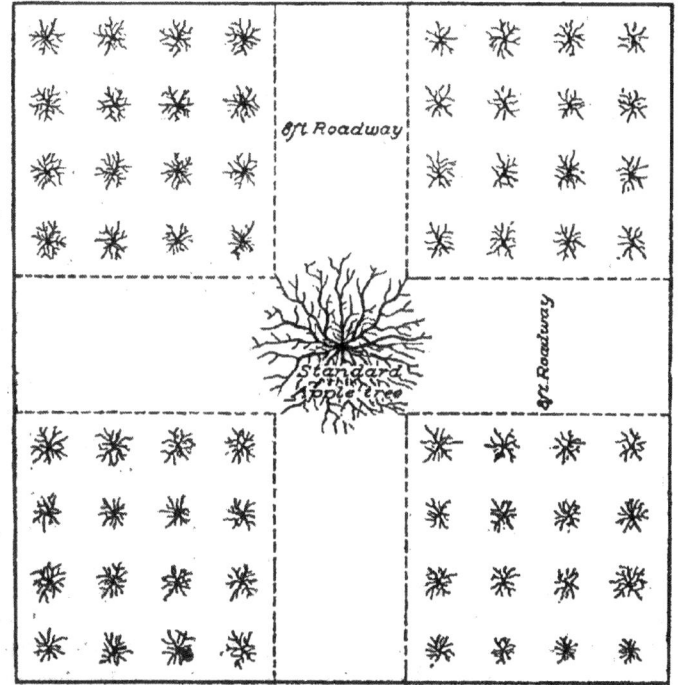

Standard Apple Trees planted 40 feet apart require 1600 feet area and begin to bear in 6 or 8 years. While 64 Bush Trees may be planted on same 1600 feet and begin bearing the 2d year from planting. After Thornton.

Fig. 58

for comparison, Paradise trees at 4 feet apart. That is, the standard apple trees occupies a space of 1,600 square feet and the Paradise 16 feet. Or, theoretically, 100 Paradise trees will fit on the same area of ground as one standard apple tree; this is, however, impracticable, as roadways must be provided for attending to the trees properly (see cut). We will therefore be reasonable and plant four rows four feet apart on each side of an eight-foot roadway, making 40 feet each row would then contain eight trees, making 64 trees for the block of 40 feet square and leaving an eight-foot roadway through the middle of the plot; this would be equal to 1,628 trees to the acre, not counting fractions, with proportionate roadway space. A little practical figuring upon the basis of the foregoing table of the actual yield of these little trees will give some startling results. We find that one of these bush trees yielded in six years over five bushels of apples, or 320 bushels from the trees occupying the space of one standard apple tree THAT HAD NOT YET REACHED THE BEARING AGE; consequently as only 27 apple trees, at 40 feet apart, fit on one acre, 8,640 bushels of apples could be produced from one acre (not counting fractions) of dwarf apple trees in six years from planting and BEFORE ONE ACRE OF STANDARDS PLANTED AT THE SAME TIME HAD COME INTO BEARING. These dwarf trees would continue bearing increasing quantities of fruit for 15 or 20 years longer, when the orchardist could well afford to dig them up and plant fresh. The generality of standard apple trees require five or six years to BEGIN bearing; the bush, on the contrary, beginning to bear the second year and steadily increasing its crop till 10 years old and continuing to yield steadily maximum crops till 20 or 25 years old. On the other hand the standard tree may be expected to gradually increase till from 15 to 20 years it will yield about three barrels (nine bushels) per tree per year; from this time until the trees begin to fail from old age, the annual yield will be under 15 bushels. These are average figures for well cared-for trees, and allowing for off years, poor years, and poor trees, or badly pruned or moss-bound trees, will not do so well. The standard tree will bear for 50 years, and it will average for good and bad years 10 bushels a year, or 500 bushels in all. This, I think, is a liberal estimate for one standard tree on 40 feet square of land. Compare the above with the Paradise apple, or, rather, 64 of them, for that number of trees may be grown on the same area of land as one standard tree. One dwarf Paradise apple bush, as we have seen, will bear at an average 1½ bushels a year, or 31½ bushels, which multiplied by 64 (the number of trees on the 40 feet square allowed for the standard tree) will give 2,400 bushels Allow margins to suit yourself. Of course we will have the same standard tree bearing for another 25 years, but we will only have to wait for four years to plant another lot of 64 trees, and have them catch up and pass the old standard and repeat the experience which we can very well afford to do. But this is not all, for it must be remembered that the dwarf apples are superior in size, beauty, quality and selling price to those grown on standard trees, and every apple on these little bushes is within reach of one's hand from the ground and may be thinned without difficulty; there are no windfalls to amount to anything; the work of spraying is reduced to a minimum. You all know what a tiresome job it is gazing up to the sky looking for tent caterpillars' eggs on a 40-foot tree. while the same is only pastime on those small bushes. No packing of awkward ladders in pruning time, or climbing trees in picking time Of course 64 trees require more attention than one and cost more for a start, but the work is light in character and such as any boy, girl or woman can do, and most of it a real pleasure. Anyhow, who would begrudge the work when the returns are so liberal. In the above comparison my remarks referred to the dwarf apple in the bush form, and as profitable even as it shows up with bushes at four feet apart or occupying 16 square feet each, how much better results may be expected if we use cordons either upright or oblique or U form, which may be planted in rows four to six feet apart and only two feet apart in the row, occupying eight to twelve feet in area, and yet are individually as productive as the bushes—or nearly so. In this case, instead of 64 trees occupying the area of one standard tree, we would have 96; or instead of 1,628 bush trees per acre, we would have 4,224 oblique or upright cordons. These figures may be astounding, but no more than if we compare the old stage coach with railway trains of the present day, or comparing the old-fashioned plough with the up-to-date steam plough, or the reaping hook with our best harvesters. Our little dwarf fruit trees offer the same gigantic stride in advance in the horticultural field, combined with intensive culture. Nor is there anything visionary in the statements, as they are established facts, though not yet exploited to the same degree, but will be in the near future, when we secure the irresistable combination of grit, money and knowledge duly harnessed Of course intensive development of the dwarf fruit tree idea is more costly both to start in the outset and in its subsequent exploitation than the ordinary commercial orchard, but in view of the enormously increased returns of profit this condition cuts no figure as all advances in industrial lines are subject to the same condition There is, however, in this horticultural advance the great advantage that it works as well on the limited area of the 40 feet square, or the acre, as it does on the 100 acres, being merely a question of capital The labor question which has been a bugbear to the horticulturist, may be worth consideration, but is by no means difficult to any one who bases his every day business on the golden rule. The increase of labor required under this system is really very insignificant as compared with the returns, and will follow the industrial experience in other lines of industry. The present supply of labor is inadequate because the pay and treatment of labor is unsatisfactory to the laborer. When the typesetter was introduced, many worthless or indifferent printers lost their job, but the better class were promoted to the machines with better pay. So with other industries, but great adverse influences existed The captains of industry put up the prices to "all the traffic would bear," while they cut the wages to all the laborer would bear; on the other hand, the trades union is started in "to kill the goose that laid the golden eggs" by restrictive regulations, etc , both parties being antagonistic to the golden rule Under present conditions the margin of profit to the orchardist is so narrow and the supply of RELIABLE LABOR so scanty, the question, How to help matters? is pressing, the answer will be found in adopting the dwarf fruit tree culture, which will enable us to so materially improve the workman's condition, that he will so speak, quit the union and "paddle his own canoe " In this connection the story of the Westinghouse Company's experience in England is very instructive. The Westinghouse Company is a celebrated American firm of world-wide reputation They had occasion to establish a gigantic factory in England, one requiring the use of several million bricks in its construction. At the start they found themselves "up against" the builders' trade union, who would not permit their bricklayers to lay more than 400 bricks in one day The Westinghouse Company with American strenuosity nevertheless started the work, not only giving higher wages than other builders were giving, but added premiums and bonuses in accordance with the quality of the work performed, giving the shirkers and inefficient men immediate discharge. The result was, as might have been expected, the union tried bluffing and boycott, but it would not work, and the men finding they were treated with strict justice and liberality, ignored the union. Good men were favored and drones hunted; walking delegates were not admitted to the works; the good men put in their best work, and from laying only 400 bricks a day, as at the start, soon achieved the laying of 900 bricks per day as the ordinary day's work. Let us see now in what position the commercial orchardist stands to meet the labor question, depending entirely on standard trees or substituting in whole or in part the dwarf trees. We have seen that a standard apple tree will average for 50 years 10 bushels a year, and the cash returns will be less than an average of 50 cents a bushel, or $4 per tree, 27 trees to the acre gives $108 per acre. With dwarf trees we have 64 bushes or 96 cordons on 40 feet square of land that will yield for 25 years an average of 96 to 144 bushels per year from the same 40 feet square of land. While the standard apples averaged 40 cents a bushel, these dwarf apples, being so much superior in appearance and quality will reach an average of $1 50 or more per bushel, consequently will return $144 to $216 per 40 feet square, or multiplied by 27, will net $3,888 to $5,892 per acre. Again strike your own margin. With such a showing we could afford to give our workmen a rate of wages beyond their wildest imagination, and steal the labor union's thunder. The very best of men would be tumbling over one another in competition to secure such employment, and when they were lucky enough to obtain it would shrink from no effort to retain it permanently.

OF DWARF FRUIT TREE CULTURE.

Commercial Orchard. One acre with 40 standard Apple Trees 33 ft. apart and 486 Bush Trees as fillers
Fig. 59

Fig. 59 shows another style of orchard planting, where dwarf trees are used as "fillers" in COMBINATION WITH standards In this place standard apple trees are planted two rods apart (33 feet), which allows 40 standards to the acre, and in ADDITION 480 dwarfs on Paradise stock are planted eight feet apart, as shown. In this plan each 33 feet square is supposed to be divided into quarters and three dwarfs planted in each square, omitting the corner next to the standard, leaving them without a dwarf On this place no roads are repaired as the trees are eight feet apart, which allows carts and sprayers to be freely moved anywhere among the trees.

Now just here comes the progressive American with iconoclastic tendencies (that is me), and looking over the plan, says: "What is the good of these standard trees, anyhow? Why not dig them out and fill their places with 160 additional dwarfs, making 640 trees to the acre?" By so doing we will be changing the dwarf system from a "COMBINATION AS FILLERS" into a "DIRECT COMPETITION" with the standard plantation. The standards are a nuisance anyway, requiring intolerable labor and cost for pruning, spraying, pest fighting, thinning and harvesting the fruit, not to mention the waste from windfalls, overbearing and the impossibility of complete protection from infectious diseases and insect enemies, as well as the long years of delay in waiting for them to reach a profitable stage of production and the lower grade in size, quality, beauty and market value of the fruit produced, as compared with the dwarf trees These latter have practically no loss of fruit from windfalls, and all the cultural manipulations, while requiring to be performed with due care and at the proper time, are so much reduced in laborousness as really to be classed as a pastime and interesting occupation, but above all the early maturity, large size, high quality, beauty and prolificacy as well as the higher market price of the fruit, raises them far above comparison with the effete standards

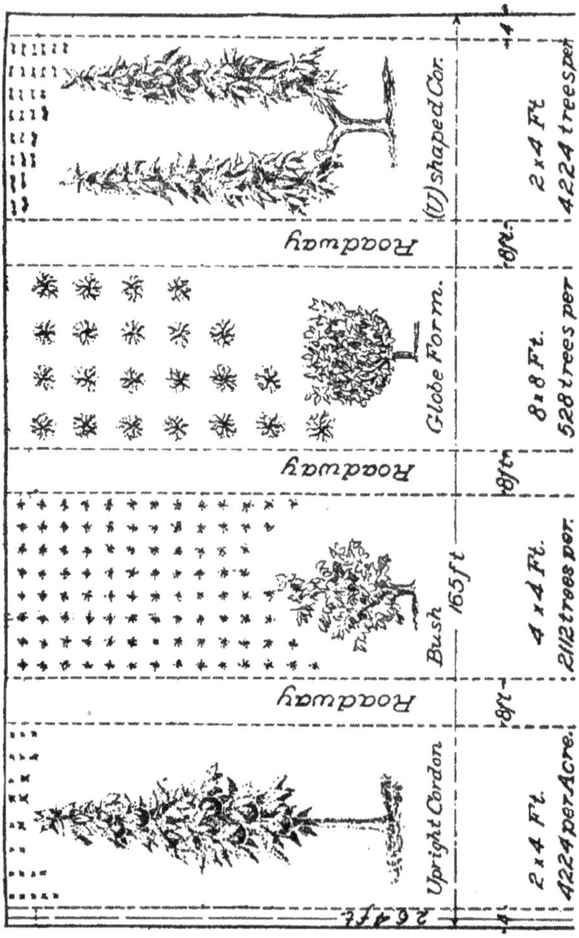

OF DWARF FRUIT TREE CULTURE. 71

One acre, 165x264 ft. (cut short to fit page) showing 4 styles of planting Dwarf Trees in COMPETITION with commercial orchards. After Thornton

Fig. 60

'Having thus introduced the subject of DIRECT COMPETITION as a stage in advance of mere "CONNECTION AS FILLERS," let us look into its capabilities a little more in detail. Fig. 60 shows a COMPETITIVE plan for laying out a dwarf tree orchard or garden. Here we take a piece of suitable land, 165x264 feet, which equals one acre of 43,560 feet, and lay out three eight-foot roadways, as shown, dividing the plot into four one-quarter acre lots, each being 41¼ feet wide and 264 feet long. Right here comes the first question to be decided regarding the style of trees to be planted and the distance apart. We have bushes at four by four feet apart, and oblique, upright and U shaped cordons to select from. Let us compare the merits of each. With bushes at four by four feet apart we could plant eight rows in each quarter acre, with 66 bushes in each row. That would allow 528 bushes to the quarter acre, or 2,112 bushes to the full acre. These would come into bearing the second year, and by the fourth year one bushel per bush might fairly be expected; that would be about 5½ bushels in the first four years from planting, or 11,616 bushels per acre, and observe here that an orchard of standard trees, planted at the same time, would hardly show a solitary apple, although there might be 40 of them on the acre.

Now let us consider oblique and upright cordons, which amount to about the same, the oblique having 25 per cent more bearing wood than the upright cordons. The cordons may be planted two by four feet apart. There would still be eight rows four feet apart in each quarter acre, but only two feet apart in the rows, or 132 cordons in each row; we would thus have 4,224 cordons to the acre, and would practically bear the same quantity as the bushes above described; or 4,224 bushels the fourth year, or 23,232 bushels in the first four years.

This is almost beyond belief, but is merely the result of intensive culture, a system as yet only in its infancy. Wonderful as the above results may appear, the next plan will double up the yield Instead of planting bushes or plain cordons we advance a step and plant U shaped cordons (see Fig. 60). We just double the bearing wood on the same number of trees. U cordons being planted at two by four feet apart This last proposition I leave you to figure out for yourselves, and yet that does not reach the limit by any means, for we may further intensify intensive culture by planting those cordons at only 18 inches apart in the row, which may be safely accomplished But I must stop here or my readers will think I am "giving them a pipe dream." I will therefore only repeat here the words of St. Paul when Festus said, "Paul, thou art beside thyself. Much learning doth make thee mad." He replied, "I am not mad, most noble Festus, but I speak forth words of truth and soberness "

Fig. 60 illustrates this very clearly. It is meant to represent a portion of one acre (cut off to fit on the page.) It represents 165 feet wide and if carried out to full length 264 feet long, in it are the three eight-foot roadways, thus dividing into four one-quarter acre tracts, on the left is shown a tract planted in upright cordons at two by four feet apart of 4,224 trees per acre Next comes a tract devoted to bushes at four by four feet apart, or 2,112 trees to the acre Next we have a tract with globe or goblet form bushes, which are a little more spreading and as set at eight by eight feet apart, or 528 trees per acre; and on the extreme right we have the U form cordon with the same number of trees as the upright cordons, and occupying the same space, BUT WITH JUST TWICE THE AMOUNT OF BEARING WOOD; and if need be, you can intensify this intensive culture by 25 per cent by planting those upright cordons 18 inches apart in the row instead of two feet.

Naturally it may be asked here, "If these facts have been known in Europe for ages, why have they not been commercially exploited there?" Well, it is easy to ask questions but not always so simple a matter to answer them satisfactorily. I will, however, give a scrap of history connected with an analogous case that may point to the answer. Over one hundred years ago there was a man who went out to play with his boys and show them how to fly a kite, when a thunder storm came along and the kite string got wet and thus became a good conductor of electricity, and he found the current of electricity was conveyed from the cloud to earth along the wet string This was Benjamin Franklin. The fact was established and duly proven, but remained unutilized for many years, when another man with his head screwed on differently came along and viewing the conditions, brought his imagination to bear and said, "Why can't we stretch insulated wires to conduct the electricity from place to place and utilize it where required?" and he stretched his wires and sent the celebrated message from Baltimore to Washington, "What hath God wrought?" This was Morse with his knowledge, imagination and faith, and since that time we have had the electric telegraph in operation and exploited almost to the limit of possibility. When another man comes along with more knowledge, more faith and greater imagination, with head screwed on in an opposite direction to that of Morse and says "Why not take those wires away altogether and send our electricity to make it own path through space?" This was a superhuman effort, and this was the immortal Marconi. He established a dispatching station at one side of the Atlantic ocean and a receiving station on the other side, with 3,000 miles of ocean between, having his skilled assistant at the dispatching station with instructions to keep sending a message consisting of the crooked little letter S and keep on till he received further instruction, while he, the immortal Marconi, stationed himself at the other station to await results. At the time appointed both were on duty, when Marconi felt (if he did not fully realize it) that there was some influence being exercised on his instrument, indefinite, uncertain, but as the dispatcher kept on repeating the letter S its symbol flickered and wavered till at last the finger of God traced that letter S in the sight of Marconi as clearly and distinctly as that same finger of God in long ages gone by traced the fatal MENE MENE TEKEL UPHARSIN on the walls of Balschazer's banquet hall, and we had the wireless telegraph an established fact through the knowledge, faith, and above all, the imagination of Marconi under God's supreme blessings.

Now compare our dwarf fruit question with the above scrap of history and note the resemblance. None of those developments added anything to the inherent powers of electricity They already existed from the foundation of the world, but simply were unrecognized, and the men their heads screwed on in the right direction to see the glorious vista spread before them and the imagination to appreciate it had not yet come. So with the dwarf fruit tree question. More than 1,000 years ago the Japanese gardeners became aware of the possibilities of dwarfing fruit trees. Hundreds of years later the system was practiced in Europe Fruit growing was practiced from the days of the Garden of Eden, and the industry grew up in the long courses of the ages, line upon line, precept upon precept, here a little and there a little, till we reached the present state of commercial orchardizing When a man comes along with his head screwed on in the proper direction to see the glorious vista opening before him of the future of dwarf fruit tree culture and blessed with the imagination to realize it in all its detail and practical knowledge of the subject, and although 75 years of age, with mental activity sufficient to carry out his investigation in spite of the silly vaporing of Dr. Ossler, who thinks men should be narcoticized with eternal sleep at 50 years of age. And now I have personal knowledge of the various stages and advances of fruit culture from the planting of an apple or pear bud in the corner of a fence to the advanced intensive orchard culture with both standard and "fillers." When the comparison between the standard and the dwarf fruit trees in actual and direct competition comes up, I give my decision in favor of the dwarf trees every time for those who will make them a hobby.

Among the simple questions that are hard to answer I am here reminded of one in connection with fruit trees that is very curious. Twenty-five years ago the apple tree tent caterpillar used to lay their eggs in a circle around the terminal twigs and after the leaves fell they were clearly visible to the naked eye and easily removed. and in some localities the orchardist was in the habit of removing them by bucketsfull and burning them, and not one lot in 10,000 would be placed otherwise than in a circle as above. In the course of years, however, as the fight became more strenuous between the orchardist and the moth, Mrs Moth learned the trick of plastering the eggs in a flat layer on the upper side of the branch, where they were invisible to the orchardist from the ground. Any experienced and observant orchardist will corroborate this statement. Simple question: How did Mrs Moth learn this trick? I do not know, unless Mrs Moth in some way became acquainted with Whitcomb Riley's celebrated refrain,

THE GOBLIN
    WILL GIT YE
        IF YE DON'T
            WATCH OUT!

This discussion on the adaptability of dwarf fruit trees to the uses of the commercial orchardist, either in connection with standard trees as fillers or direct competition with them under intensive culture may be epitomized with advantage as follows:

Fruit trees have from the beginning been subject to certain laws and conditions and the ignorance of such laws and conditions throughout the ages in no way justifies the denial of their existence. Just as in the case of electricity, the inherent power of electricity existed from the beginning, though Morse and Marconi did not come to exploit them till the Nineteenth Century—nevertheless they always existed.

Many discoveries have been made regarding dwarf fruit trees that are now established facts fully proven and undisputable. Many other facts have not yet been fully demonstrated, but that is no reason to claim they are fallacious, but merely that they require further investigation and practical experiment to fully develop their full force.

Of the many facts already fully established and beyond cavil are the following:

First—Ordinary fruit trees are susceptible under certain treatment to the dwarfing process

Second—The dwarfing process has the power of reducing the size of the trees, so that they may be planted at distances of 9x9, 8x8, 4x4, 2x4 feet apart, and even less.

Third—That the dwarfing of fruit trees hastens their maturity, causing them to come into bearing in two years, and frequently the first year from the bud or graft.

Fourth—Dwarfing also has the effect of increasing the yield of fruit, enlarging and beautifying the fruit both in color and quilty and enlarging its size.

Fifth—From the small size of the trees and their adaptability to training in various fancy forms they are especially adapted for use in suburban lots or small patches.

In addition to this all the cultured manipulation from the nature of the case are reduced to a minimum, and such operation as training, thinning the fruit, spraying, destroying insects, pests and diseases, gathering the fruit and pruning, can all be done while standing on the ground without the use of ladders or climbing the trees.

These are all well established facts and fully settle the question of adaptability for suburbanite's use. The above facts also settle the question of their adaptability for use in commercial orchards as "fillers" to secure early and profitable crops of fruit while waiting for the large standard trees to come into bearing.

When we come, however, to the question of full competition with the large standard trees there are some facts and data that require further elucidation, not because favorable conditions do not exist, but merely that we have not yet exploited them sufficiently for practical purposes

Among those questions the most important probably is the securing reliable data of the yield of dwarf trees when planted "en bloc" by the acre for commercial purposes. As I am not aware of any extensive experiments having been made to settle this question and consequently a reasonable conservative caution would teach the wisdom of going slowly at this stage, but at the same time the acknowledged merits of those dwarf trees are amply sufficient to justify any progressive orchardist in testing the question on a limited area for his own satisfaction, and I am now planning a series of practical tests to secure reliable data upon this very important phase of the subject and I expect the results will far surpass the wildest imagination of the most optimistic orchardist, for no one has yet reached the limit of the results possible to obtain from intensive culture of any crop Many years ago Orange Judd of the American Agriculturist gave a prize for the largest crop of potatoes to be grown on one acre, and if I recollect rightly, I think there were 720 bushels from one measured acre in the prize crop, while the ordinary potato crop for the United States does not reach to more than 100 to 150 bushels per acre, and I expect to see in the near future (when the dwarf fruit trees come into actual competition with the old and effete style of standard trees) the experience of the potato grower far surpassed by the up-to-date dwarf fruit tree orchardist

In the foregoing I think I have made a fair comparison and have been fairly conservative in my figures, and, I trust, have made the subject sufficiently plain for the reader to arrive at an intelligent idea on the subject I think I have shown sufficiently valid reasons to justify an unprejudiced trial of the two systems subject to your own conditions Far be it from me to recommend you to rush into this work wildly, but go to work conservatively and try a few dwarf trees to make sure you are right and then go ahead for all there is in it. I will tell you frankly at the first word that if you are a slack handed fruit grower you had better let dwarf fruit trees alone, but if on the other hand you will take an interest in the work you will soon regard these little bushes as little pets, and watch their progress and development under your guilding care, and will grow fonder and prouder of them year by year. In such case they will amply repay all your efforts and prove a grand success outside of any pecuniary return. So far I have considered the two systems as opposed to one another and have not touched on the combination of the two. This is a very important phase of the question and worthy of careful consideration We know that in starting a commercial orchard of standard trees we require to wait five or ten years to reach the bearing age, but what are we to do for a profitable return in money from the land in the meantime? The practice has been to plant potatoes, corn or other crops between the trees, but this is not always advisable as the crops rob the young trees at the very time they require all the nourishment within their reach. Here comes in the advantage of dwarf trees as "fillers," being planted at the same time as the standards they begin to bear at two years and yield profitable crops continuously long before the standards yield any return. The doucin and crab stock are most suitable for this purpose, and may be trained as half standards. The mere mention of this subject will be enough to draw the orchardist's attention to its value and importance.

The commercial orchardist, while familiar with ordinary fruit trees, may not have had his attention drawn to this subject of dwarf trees, and may be desirous of more detailed information. I would refer them to the first part of this handbook.

It has been found by long experience that some varieties the different fruits respond better to the dwarfing process with the result of producing a far higher quality of fruit than others, consequently the European experts have made lists of selected varieties of fruit that will afford the greatest satisfaction to the grower. One of these lists I append. There are many other varieties which may be substituted for them without much disadvantage, but one must draw the line somewhere and the high reputation of this list has been established by good judges

LIST OF TWENTY-EIGHT BEST APPLES FOR DWARFING.

(In their order of ripening.)

(C) for Culinary. (D) for Dessert

Summer Apples.

MR. GLADSTONE—August (D).
BEAUTY OF BATH—August (C).
IRISH PEACH—August (D).
DEVONSHIRE QUARENDEN — August (D).
KESWICK CODLIN—August and September (C).

Autumn Apples.

POTT'S SEEDLING—September (C).
EMPEROR ALEXANDER—September to November.
CELINI—October and November (C).
STIRLING CASTLE—October and November (C).
ECHLINVILLE SEEDLING—October to January (C).
WORCESTER PEARMAIN—August and September (D).
KING OF PIPPINS—October to January (D).
COX'S ORANGE PIPPIN — October to February (D).
PEASGOOD NON-SUCH — October to January (D).

WARNER'S KING—October to January (C).
RED ASTRACAN—August and September (D).
HAWTHORNDEN—October to January (C).

Winter Apples.

GOLDEN NOBLE—October to March (C).
BISMARCK—November to February (C).
BRAMLEY'S SEEDLING—December to March (C).
GASCOIGNE SEEDLING—November to March (C).
ALFRISTON—November to April (C).
NEWTON WONDER—November to May (C).
BLENHEIM ORANGE—November to February (D).
MANNINGTON'S PEARMAIN—November to March (D).
CLAYGATE PEARMAIN—November to March (D).
COURT PENDU PLAT—December to May (D).
CORNISH GILLIFLOWER—December to April
DUKE OF DEVONSHIRE—December to May (D).

BEST PEARS.

Early Summer Pears.

BEURRE GIFFARD—End of July.
CLAPP'S FAVORITE—August.
JARGONELLE—July (Wall).
BARTLETT—August and September.

Autumn Pears.

DURONDEAU (DE TONGERS—October.
LOUISE BONNE DE JERSEY—October.
PITMASTON DUCHESSE—October and November.

Late Summer Pears.

BEURRE D'AMINLIS—September.
MARGUERITE MARILLAT—September.
JERSEY GRATIOLI—September and October.
BEURRE SUPERFIN—September and October.

Winter Pears

DOYENNE DU COMICE—November.
BUERRE DIEL (ROYAL)—November and December.
MARECHAL DE LA COUR—October and November.
MARIE LOUISE—October and November.
JERSEY CHAUMONTEL, November and January.
OLIVIER DE SERRES—March.
JOSEPHINE DE MALINES—January to March.

Best Baking Pears

BELLE DE JERSEY—November to May.
CATILLAC—December to March.

BEST SIX PEACHES.

EARLY ALEXANDER—July.
NOBLESSE—August and September.
ROYAL GEORGE—August and Sept.
HALE'S EARLY—July.
GROSSE MIGNONNE—September.
PRINCESS OF WALES—September.

BEST SIX NECTARINES.

EARLY RIVERS—August.
ADVANCE—August.
LORD NAPIER—August.
STANWICK ELRUGE—End of August.
PITMASTON ORANGE—September.
PINEAPPLE—September.

BEST APRICOTS.

MOORPARK.   HEMSKIRK.   ROYAL.

BEST PLUMS.

KIRKE'S PLUM.
VICTORIA.
GREEN GAGE.
MONARCH.
GOLDEN DROP.
POND'S SEEDLING.
JEFFERSON.

BEST CHERRIES.

BLACK BIGARREAU.
EARLY RIVERS.
MAY DUKE.
WHITE HEART
ROYAL ANNE.
BLACK TARTARIAN.
STRANG LOGIE.
NOBLE.

FIGS.

BROWN TURKEY.   WHITE MARSEILLES.

GOOSEBERRIES.

CROWN BOB.
LANCASHIRE LAD.
RIFLEMAN.
WARRINGTON.
WHINHAM'S INDUSTRY.
GOLDEN DROP (or Early Sulphur).
VICTORIA.
KEEPSAKE.
MAY DUKE.
WHITE SMITH.

CURRANTS.

COMET (New).
RED DUTCH.
RABY CASTLE.
RED VERSAILLAISE.
FAY'S PROLIFIC.
WHITE DUTCH.
BLACK NAPIES.
BLACK CHAMPION.
LEE'S PROLIFIC.

The following list of selected dozens of apples for special qualities may be of interest to persons wishing to plant choice varieties for exhbition purposes:

Large Size.

ALFRISTON.
BISMARCK.
BRAMLEY SEEDLING.
ECKLINVILLE SEEDLING.
EMPEROR ALEXANDER.
GLORIA MUNDI.

LORD SUFFIELD.
MERE DE MENAGE.
MONSTREUSE INCOMPARABLE.
PEASGOOD'S NON-SUCH.
POTT'S SEEDLING.
WARNER'S KING.

Bright Color.

BISMARCK.
CELINEE.
COX'S POMONA.
DEVONSHIRE QUARENDEN.
EMPEROR ALEXANDRE.
GASCOIGNE'S SCARLET.

HOLLANDBURY'S ADMIRABLE.
LADY HENNIKER.
MERE DE MENAGE.
RED ASTRACHAN.
THE QUEEN.
WORCESTER PEARMAIN.

Fine Flavor.

ALLINGTON.
BLENHEIM ORANGE.
CORNISH GILLIFLOWER.
COX'S ORANGE PIPPIN.
DUKE OF DEVONSHIRE.
GOLDEN PIPPIN.

IRISH PEACH.
KING OF THE PIPPINS.
MARGIL.
MR. GLADSTONE.
RIBSTONE PIPPIN.
ROYAL RUSSSET.

Heavy Crops.

ALFRISTON.
BISMARCK.
CELINEE.
DEVONSHIRE QUARENDEN.
ECKLINVILLE SEEDLING.
HAWTHORNDEN.

KESWICK CODLIN.
LANE'S PRINCE ALBERT.
LORD SUFFIELD.
POTT'S SEEDLING.
STIRLING CASTLE.
WORCESTER PEARMAIN.

I will here make an extract from Mr. P. Le Cornu's work on cordon fruit trees that may be of interest. The cordon system of growing fruit trees as adopted in the Royal Garden at Sandringham Palace, is now becoming very popular, and deservedly so, for by no other means can the same quantity of fine, highly flavored fruit be produced in any given space. Apples, pears and plums succeed as cordon, but more especially the former. Pears are also very profitable when grown in this manner and produce an abundance of fruit of larger size and better quality than that which is grown on pyramids, or ordinary wall trees. For the following reasons I hold that this is the best of all systems and firmly believe it would be adopted by many more if they only knew the advantages which are to be derived from it.

Fruit of the Largest Size and Quality—Only one rod or stem having to be supported, all the fruit borne is of the largest size and best quality.

Wall or Espalier Covered in a Short Time—A wall or espalier can be covered with trees in less than a third of the time occupied in covering it with fan-trained or other trees.

Trees Bear Younger and Give Heavier Crops—The trees treated in this way turn to bearing much younger and produce double the crops which could be expected from a single tree, covering the same space.

Upright Cordon Apples, 2 x 4 ft. apart, in bearing—After Le Cornu
Fig. 61

Walls Never Entirely Bare—If one tree dies it can easily be replaced, whereas with a fan-trained or other large tree part of the wall is left entirely uncovered for years.

Summary of Reasons—To sum up in a few words. By no other means can trees be so quickly made fruitful. The second season most of the cordons paid the cost of their purchase many times over.

THESE REASONS SPEAK FOR THEMSELVES.

He says further, in speaking of his profitable cordon fruit garden (See cut above): "The walls surrounding the illustration of the profitable garden, show clearly the leading features of this system Single cordon trees with a quantity of fruit spurs already on them should be procured of apples on Paradise stock and pears on the quince. These should be planted at an angle of 40 degrees to 50 degrees, according to the height of the wall, about 16 to 18 inches apart from one another. IN NO CASE SHOULD THEY BE PLANTED AT A GREATER DISTANCE, as the roots would have too much room for development and would cause the trees to run to growth instead of forming fruit spurs I find it more convenient to stretch horizontal wires along the walls at about every foot instead of tying in the trees with nails in the old-fashioned way, taking care to keep the wire three inches away from the wall, so that the spur at the back of the stem of the cordon may have room to develop Fruit growers who are not the happy possessors of walls need have no difficulty in growing large fruit without this expensive adjunct, for with the cordon system on wires magnificent apples and pears may easily be grown.

The lines of wire are made fast to terminal pillars, five to seven feet high at each end with intermediate pillars at every ten or fifteen feet, the whole being tightened by means of raidisseurs or stiffeners. The pillars may be made of wood or iron If the former, they should be made like an inverted cross and tarred or painted to preserve them Iron, owing to its lasting properties, is really the cheapest in the end. All my pillars formerly were of wood, but have now been entirely replaced by iron work After much thought to the subject, I have adopted the system here illustrated, which for rigidity cannot be beaten (See cut No 27). When convenient the rows of cordons should be placed north and south, so that the sun may ripen the fruit on both sides of the trees.

The Horizontal Cordons—Are usually planted as an edging to garden paths, and in this way they make very handsome objects and occupy very little space. Double horizontal cordons occasionally become unequal in strength, hence I always recommend single cordons, planted to follow one another in one direction A line of wire should be stretched 15 or 18 inches above ground The cordons should then be planted (apples in preference) at every six or eight feet, and then be made fast to the wire in such a manner that the part which is below the wire may be perfectly perpendicular, after which the remainder of the stem should be carefully bent down and tied full length to the horizontal wire.

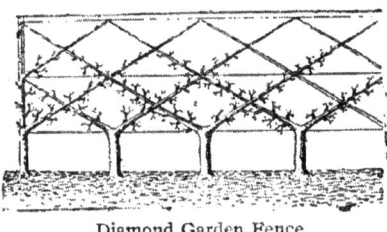

Diamond Garden Fence
Fig. 62

Diamond Fence Patterns (see garden cut)—Double cordon apples should be planted for this purpose at 18 inches apart. One of the branches should be trained to the right at an angle of 45 degrees, and one to the left at the same angle, the two forming togethr a perfect right angle, and as the trees grow the leading branch should be trained in a direct line until the desired height is attained. This will form a very picturesque and in many cases a very useful fence or partition between two parts of the garden. A wire fence will be required the same as in the oblique system, and the distance between the wires should be so regulated that the line of wire may pass exactly behind the crossing of the branches, forming the corners of the diamonds.

Upright Perpendicular Cordons (for very high walls and arches) —Are recommended for arches, and when walls are at least 15 or 20 feet high and as the sap has always a tendency to flow upward it will be necessary to shorten the leader back each season in order to develop the fruit spurs along the stem. The varieties of apple, pear and plum best adapted for cordons are those that have close-eyed and short jointed wood. For cordon plums the soil should be as poor as possible. Lime rubbish and rubble of any kind may be mixed freely with the soil in planting and no manure whatever should be employed except in the poorest of soil. Lifting these occasionally will prove very beneficial.

Having given an English expert's instructions in the art of dwarfing trees, it may not be out of place to quote from a Japanese source the instructions they issue to their customers and note how

OF DWARF FRUIT TREE CULTURE.

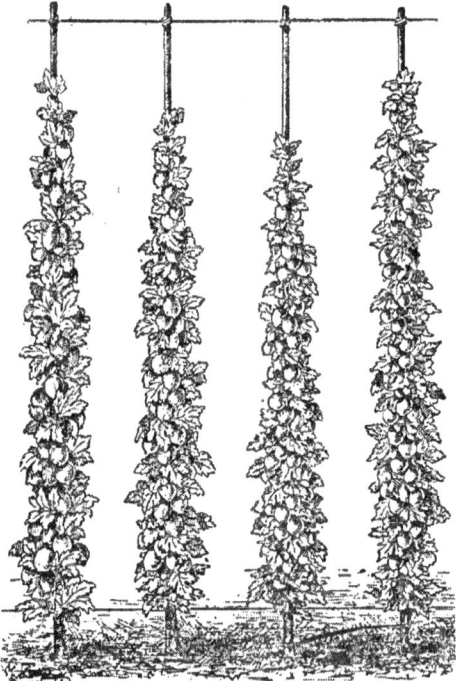

Upright Cordon Gooseberries. Only 9 inches between the plants. After Philip Le Cornu
Fig. 65

closely they agree with the English practice, although they were experts in the art of dwarfing trees centuries before the English horticulturists ever heard of the subject. To such perfection have they brought the art that dwarf trees of over 400 years growth in pots are to be seen at the present day, sound and healthy, still growing in pots.

Treatment of Thuja Obtusa (a variety of the Arbor Vite)—During spring and summer, by preference, keep this plant in a sunny, airy situation where the wind will pass freely through the branches, water once a day, giving just enough to make the soil moist; in dry, hot weather it may be necessary to give water twice a day, care, however, being taken not to have the soil wet, and never water unless the plant needs it. Watering overhead in dry weather is bad, but rain is always beneficial.

During winter keep the tree in a cool green house, partially shaded or in an unheated orangery, giving water about once in 10 days; the soil, however, must never be allowed to get dry. (The secret of successful culture of all plants in pots, consists in judicious watering, giving too much or too little is equally bad. Maples and other deciduous trees (such as fruit trees) take the same treatment as Thuja as regards watering, but are much more accommodating than evergreens. In fairly mild climates the maples may remain out-

Pear Tree, "Madam Treyve," Sept.
Goblet form with 8 branches, 10 years old, 6 ft. high, 11 ft. circumference, with 138 fruits
Fig. 66

of-doors all winter, but where the frost is very severe they should be kept in a cellar after the leaves have fallen in autumn. The soil must always be kept moist but not wet. Early in spring put the plant out of doors and fully exposed to all weathers, and when in full leaf use for decoration in doors as needed.

Manuring—When the trees commence growing in the spring, we give manure twice a month, say March, April, May and June, again in September and October. In the hot days of July and August we give no manure, and the same in winter and spring, the plants then being at rest, the best manure is finely powdered oil cake or bone meal. To a jardinier one foot in diameter we give three or four large teaspoonsful, not heaped, of this dry manure, spread evenly round the edge of the jardinier—a larger or smaller jardinier will require more or less—for a small jardinier, say three by six inches, half a teaspoonful will be ample each time.

Repotting—This is done by us once in two or three years, as follows: Lift the plant out of the jardinier and with a sharp pointed stick remove about one-third of the old soil around the edge and bottom, cutting away a portion of the old fine roots. but none of the strong roots, then replace the plant in the same jardinier, first looking to the drainage. For a small shallow jardinier we use a flat piece of stone or a flat crock over each hole; over this we spread some rich, fresh soil to within half an inch of the rims; this holds the water and prevents the manure being washed over the sides of the jardinier; also the soil should be made sufficiently tight around the edges of the jardinier to prevent the escape of the water, it being of the first importance that all the ball of soil around the plant be moistened at each watering. Should the watering of the plant at any time be neglected and the soil become quite dry, put the jardinier in a tub of water for 10 or 15 minutes—NOT LONGER—and if the injury is not too serious, the plant will recover. In the case of large plants we use hollow crocks for drainage, the same as used by growers of specimen plants. After several repottings the plant having increased in size, shift into a larger pot, but as dwarfness is the thing aimed at the smaller the shift the better. Repotting should be done in February or March, just before spring growth commences.

Pruning—To maintain dwarfness in trees, pinch back the young growth, this we usually do from April to the middle of June, and always with the finger and thumb. Flowering peach and flowering cherry, etc., we pinch back to non-flowering shoots either before or after blooming; in July and August we pinch back all young growth, leaving only four or five leaves on each shoot. Maple and other deciduous trees are pinched back in the same manner, leaving two to four leaves, as may be necessary to maintain the proper shape of the plants. Should a second growth be made the same rule is followed of pinching out the points.

It will be noticed here the great similarity between the European and the Japanese practice of dwarfing trees, and yet it must be remembered that while the European system dates back a few centuries, the Japanese goes back a millenium' or more

The Japanese dwarfs, when compared with the European dwarf potted trees, show a very distinctive difference in that the roots are in large measure above the soil and exposed to the air. This is because in addition to their instructions for potting there is one manipulation they carefully guard as a trade secret, and that is. Each year when they repot the trees they plant the tree very slightly shallower than it was the year before, and although in young trees this is hardly apparent, in the course of years it becomes emphasized and gives the tree the appearance of growing on stilts. Among the fancy forms of trained trees the Japanese gardener keeps in stock are "The Stork" (a favorite fancy figure with them), "The Turtle," "The Chicken," "The Rooster," and "The Hen," "The Junk Full Rigged," and offer customers to train trees to any design they may order Of course there is no practical advantage in these fancy forms except fun and fancy for the grower, and to enjoy that pleasure one had better exercise their own ingenuity to do that work for themselves

SPRAYING AND FRUIT PESTS

This being intended as a hand-book for instruction of suburbanites who have little or no practical experience in the details of orchard work, it would be incomplete without some reference to this very important detail.

During the past few years the fruit pests have greatly increased in number and variety and at the same time our knowledge about them and the means of combating them has also increased. The means at our disposal for this warfare, while efficient, must be used with energy and intelligence. To this end we must acquaint ourselves with the nature and habits of these enemies and must therefore classify them First we have two principal divisions. Insects and fungi. The insect pests may be divided into those that feed by biting the fruit and leaves and those that live by suction; the other division is in the form of vegetable and bacterial enemies. The biting insects are poisoned by arsenicals sprayed on the fruit, leaves and branches This spraying business is of the utmost importance, and like some political parties voting, must be done "early and often." As the result of spraying is perfectly successful in exact proportion to the care and thoroughness with which it is done, consequently it will be both labor and money wasted if performed in a slipshod manner. The spraying outfit consists of a receptacle for holding the chemicals, which should be constructed of brass, copper or wood, and the pump, which must also be made of brass or bronze, as no other metal will stand the corrosive action of the chemicals; there is also required a rubber hose and a spraying nozzle. Spraying outfits are manufactured in great variety, from the brass garden syringe and wooden stable bucket to the elaborate power machines, driven by gasoline motor engines and tanks of several hundred gallons capacity, and capable of spraying two or four trees at one time. As there are cheap sprayers on the market made of tin I will in this place add an emphasized DON'T, DON'T, ever purchase a tin spray pump, for it would rot out after the first use of Bordeaux mixture. Your spray pump MUST BE either brass or bronze. As the suburbanite will only have a few trees, and dwarfs at that, a very modest outfit will answer his requirements. The simple wooden stable bucket with brass hand sprayer is, of course, the simplest, and is fairly efficient where only a small space and low growing plants require treatment. The knapsack sprayer is a very convenient style for suburbanite's use. It is worn like a knapsack and is supplied in two styles, one with direct action pump and the other has the fluid forced through the nozzle by atmospheric pressure. Both are good, reliable implements and give satisfaction to those using them. As the foregoing require to be moved by hand from place to place, they are to some extent inconvenient. To avoid this difficulty one has been introduced on the principle of a barrel cart, which may be trundled about the garden with greater facility. It is needless here to refer to the large outfits operated by horse or motor engine power as they are unadapted to use in restricted areas. Having decided on the style of tank for holding and carrying the fluid, we come to consider the style of pump, and here I may say there is no "BEST" pump. All that are now on the market are capable of doing fair work; that pump is the most useful that throws the fluid with the greatest force and with the expenditure of the least labor. It is the force with which the fluid is driven through the nozzle that secures the fineness of the spray, which should be like a cloud or mist. We now come to a very important element of the spraying outfit, that is the nozzle. There are nozzles and nozzles (ad infinitum). As some nozzles are liable to become clogged with little grains of lime, a provision consisting of a movable pin has been added in order to clear away any obstruction. Of this kind is the Bordeau nozzle. The Vermorel is a very favorite pattern, but so many changes have been made (some useful, others of no importance) that one is compelled to fall back on first principles Any nozzle in which the fluid enters the nozzle chamber at one side and whirls around right angles to the outlet hole, before being forced out, should do the work Having procured our outfit we are up against the question: What are we to do with it? We had better here fall back on the first principles. We must use it with energy, in the right manner and at the right time. Remember that ONE thorough spraying when required is worth a dozen careless attempts. As many of the materials used in fighting fruit pests are highly poisonous the greatest care should be taken to keep all substances used for spraying where they will be safe from animals, children and meddlers. And all such materials should be correctly labelled.

Solutions and mixtures contain copper sulphate, corrosive sublimate and arsenate of lead, should be made in wood, glass or earthen vessels.

Arsenical sprays should not be applied to fruits within two weeks of the time they are to be used as food.

Trees should not be sprayed when they are in blossom.

Familiarize yourself with the habits and appearance of the various fruit pests and the best treatment for their eradication.

FUNGICIDES AND INSECTICIDES.

These consist of quite a variety of mixtures, some used in liquid form as sprays, come in dry form as powders by dusting, some in gaseous form in fumigations, and some combined fungicides and insecticides, so as, if possible, to kill two enemies with one shot.

Bordeaux Mixture—Is the first and one of the best fungicides adopted for controlling fungus diseases. It has long been known that the various salts of copper were destructive to fungus spores, and sulphate of copper was first used in France to control the Phylloxera, or grape fungus. It was found, however, that the sulphate (or Bluestone) contained so much free acid that it injured the foliage and consequently something was required to naturalize the acid; this was effected by the use of quicklime, and after experimenting the vinyardists succeeded in making a mixture of carbonate of copper (Bluestone or Blue Vitroil) slaked lime and water, and this became known as BORDEAUX MIXTURE. When the fungus fruit pests began to be unbearable it was introduced into our orchards to fight the pests, with varying results, as might be expected. The concensus of expert opinion, however, was largely to the effect that it possessed real merit, and that where it failed or was partially unsatisfactory was finally traced to preventable causes. After long years of practical and scientific work it has now been brought to a degree of perfection and universal application as to render its use and result therefrom absolutely certain of success if used vigorously, and with reasonable intelligence.

FORMULA.

Sulphate of Copper (Blue Vitriol) four pounds.
Lime (unslaked) four pounds.
Water, 25 to 50 gallons.

Dissolve the bluestone in hot or cold water, using a wood or earthen vessel, and hanging the bluestone tied in a cloth on the surface of the water. Slake the lime in a tub, adding water cautiously, and only sufficient to insure thorough slaking. After thoroughly slaking, more water may be added and stirred in until it has the consistency of thick cream. When both are cold, dilute each to the required strength and pour both together into a separate vessel and thoroughly mix. Before using, strain through a fine sieve or gunny sack. This seems to be a very simple matter; yet considerable trouble has freequently been experienced in the prparation of Bordeau mixture. Care should be taken that the lime is of good quality and well burned and has not been air slaked. Lumps are far better than fine lime, and are selected by masons for fine work. When small amounts of lime are to be slaked it is advisable to use hot water Lime should not be allowed to become dry in slaking, neither should it be allowed to be completely submerged in water. Lime slakes best when supplied with just enough water to develop a large amount of heat, which renders the process active. If the amount of lime in the Bordeau mixture is not sufficient to neutralize the acid, there is danger of burning the tender foliage. There are two simple tests that will show this condition, one is to dip the polished blade of a knife in the mixture. If the amount of lime is insufficient, a thin coat of copper will be deposited on the knife The other test is made by dissolving Ferro cyandie of potassium in water (one ounce Ferro cyanide to five or six ounces water), a deep brownish red color is imparted on adding the test to the Bordeaux mixture, and more lime should be added until neither reaction occurs A slight excess of lime is, however, desirable, and it is seldom one has to apply these tests

Several standard strengths of Bordeaux have been established and are known by the abbreviated names of the formulae, as follows:

Full strength (or 4-4-25, formula). That is four pounds copper sulphate, four pounds lime, and 25 gallons water.

Half strength (4-4-50, formula).

6-4-50, formula.

3-6-50, formula.

2-2-50, formula.

3-9-50, formula.

The last three formulae are suitable for peach and plum foliage, which are liable to burn when full strength mixtures are used. Bordeaux mixture is also on the market in a dry form and may be used either alone or mixed with the arsenicals and applied with a powder gun.

SODA BORDEAUX MIXTURE.

Copper sulphate, four pounds.

Water, 50 gallons.

Add enough soda lye to make the mixture alkaline to test paper. This is merely substituting soda for the lime and has the advantage of not clogging the spray nozzle, which the lime is apt to do.

AMMONIACAL COPPER CARBONATE.

Copper carbonate, five ounces.

Ammonia (26 degrees Beaume), three pints.

Water, 50 gallons.

Dissolve the copper carbonate in the ammonia. This may be kept any length of time without injury if kept in a glass stoppered bottle and can be diluted to the required strength when wanted for use. The solution loses strength on standing.

COPPER SULPHATE SOLUTION.

(Strong Solution.)

Copper sulphate, one pound.

Water, 25 gallons.

Applied only on trees without foliage.

COPPER SULPHATE SOLUTION.
(Weak Solution.)

Copper sulphate, two to four ounces.
Water, 50 gallons
For trees in foliage.

POTASSIUM SULPHIDE.

Potassium sulphide, three ounces
Water, 10 gallons.
Valuable for gooseberry mildews, etc

INSECTICIDES.
(Stomach Poisons.)

PARIS GREEN—DRY.

Paris Green, one pound.
Flour, 20 to 50 pounds.
Mix thoroughly and apply evenly, preferably when the dew is on the plants.

PARIS GREEN—WET.

Paris Green, one pound
Quicklime, one to two pounds.
Water, 200 gallons.
Slake the lime in part of the water, sprinkling in the Paris Green gradually and then add the rest of the water. For peach and other tender leaved plants use 300 gallons of water. Keep well stirred while spraying. Paris Green is a preparation of Arsenic, and a powerful poison; great care must be taken in handling it.

ARSENATE OF LIME.
(Poison.)

White Arsenic, two pounds.
Sal Soda, eight pounds
Water, two gallons
Boil till the arsenic all dissolves—about 45 minutes. Make up the water lost in boiling, and place in an earthen dish. For use take one pint of stock, two pounds of freshly slaked lime, and 45 gallons water, and spray.

ARSENATE OF LEAD.
(Poison.)

Arsenate of Soda (50 degrees strength), four ounces.
Acetate of Lead, 11 ounces
Water, 100 gallons

Put the arsenate of soda in two quarts of water, in a wooden pail, and the acetate of lead in four quarts of water in another wooden pail. When both are dissolved, mix with rest of the water. Warm water in the pails will hasten the process. For elm leaf beetle use 10 gallons instead of 100 gallons of water.

As arsenate of lead has now an established place on the market it will be cheaper and more satisfactory to procure the ready made article from the drug store.

ARSENATE OF LEAD.
(Ready Prepared Article.)

Arsenate of Lead, three pounds.
Water, 50 gallons; for coddling moth, and
Arsenate of Lead, five pounds.
Water, 50 gallons, for elm leaf beetle, and on potatoes.

CONTACT POISONS.
WHALE OIL SOAP.
(For Winter Use Only.)

Potash Whale Oil Soap, two pounds.
Hot Water, one gallon.

WHALE OIL SOAP.
(For Summer Use.)

Potash Whale Oil Soap, one pound.
Hot Water, six gallons.

KEROSENE EMULSION.

Hard Soap, shaved fine, half pound.
Water, two gallons.
Kerosene, two gallons

Dissolve the soap in the water, boiling hot; remove from the fire and pour it into the kerosene while hot. Churn this with a spray pump till it changes to a cream, then to a soft butter-like mass. Keep this as a stock, using one part in nine parts of water for soft bodied insects, such as plant lice, or stronger in certain cases.

RESIN LIME MIXTURE.

Pulverized Resin, five pounds
Concentrated Lye, one pound.
Fish Oil, one pint.
Water, five gallons.

Place the oil, resin and one gallon of hot water in an iron kettle and heat till the resin softens; then add the lye and stir thoroughly; now add four gallons of hot water and boil till a little will mix with cold water and give a clear amber colored liquid, add water to make up five gallons. Keep this as a stock solution.

For Use Take

Stock Solution, one gallon.
Water, 16 gallons
Milk of Lime, three gallons.
Paris Green, one-fourth pound.

This sticks well to smooth leaves and is highly recommended by some of the experiment stations.

LIME SULPHUR WASH.

Fresh Stone Lime, 20 to 22 pounds.
Flowers of Sulphur, 18 to 20 pounds
Water, 45 to 50 gallons.

Slake the lime with some of the water in a large iron kettle, sprinkling in the sulphur gradually. Start a fire under the kettle to continue the heat begun by the slaking lime, and boil till the mixture becomes a dark orange color, adding water till 35 or 40 gallons are in the kettle. Boiling will probably take from 40 minutes to an hour; stirr frequently, a successfully prepared lot should have little sediment when the boiling is finished. Strain through a fine meshed strainer, adding the rest of the water and spray while warm. This is a winter and fall wash, but cannot be used while the trees are in leaf.

CARBOLIC ACID EMULSION.

Hard Soap, shaved fine, one pound
Water, one gallon
Crude Carbolic Acid, one pint

Dissolve soap in boiling water, add the carbolic acid, and churn as for kerosene emulsion. Use one part of this with 30 parts of water.

HELLEBORE.

White Hellebore, one ounce.
Water, one to two gallons

Steep the helebore in a pint of water and gradually add the rest of the water. Hellebore may also be dusted over the plants, either pure or mixed with flour or plaster.

INSECT POWDER—PYRETHRUM.

Mix with half its bulk of flour and keep in a tight can for 24 hours; then dust over the plants.

Insect Powder, 100 grains.
Water, two gallons

Mix together and spray.

COMBINED FUNGICIDES AND INSECTICIDES.
BORDEAUX MIXTURE AND PARIS GREEN.

Paris Green, if pure, six ounces, more if necessary
Bordeaux Mixture, 50 gallons.

BORDEAUX MIXTURE AND ARSENATE OF LEAD.

With ready prepared arsenate of lead use five pounds to 50 gallons.

BORDEAUX MIXTURE AND ARSENATE OF LIME.

Arsenate of Lime (made by foregoing formula), 1½ quarts
Bordeaux Mixture, 50 gallons.

IVORY SOAP.

Ivory Soap (10-cent size), one bar.
Water, 15 gallons.

Apply warm as it thickens on cooling Recommended for rose mildew and plant lice

FUMIGANTS.
CARBON BISULPHATE.
(Explosive—Use With Caution.)

Evaporate one pound of carbon bisulphate to every 1,000 cubic feet of space This is done by pouring the bisulphate into shallow dishes placed in the upper part of the space to be fumigated, and closing everything tightly and leaving it 24 hours. Then open and ventilate for 10 minutes before entering or using anything that has been fumigated. The vapor of carbon bisulphate being heavier than air settles towards the floor. This treatment is effective for infested grain, weevily seed, clothes moths, carpet beetles, etc., or any living thing in closets, trunks or tight boxes where they may be placed for treatment. CAUTION. DO NOT USE CARBON BISULPHATE NEAR FIRE OR WHERE THERE IS MUCH HEAT, AS IT TAKES FIRE AND EXPLODES EASILY, EVEN FROM A LIGHTED PIPE OR CIGAR

HYDROCYANIC OR PRUSSIC ACID.
(Deadly Poison.)
(For Nursery Stock)

Potassic Cyanide (98 or 99 degrees)
Sulphuric Acid (1.83, sp gr commercial).
Water.

Multiply the number of cubic feet to be fumigated by 2 or .25, giving the number of grams of cyanide for the house or box, divide the answer by 28.35, giving the weight of cyanide in ounces. Take twice as many fluid ounces of acid and four times as many fluid ounces of water as was taken in ounces by weight of the cyanide. Mix the water and the acid in an earthenware or graniteware jar. Then by a loose bag and string drop the cyanide into the acid after closing tightly the place to be fumigated, with yourself safe from the fumes outside As soon as the cyanide touches the acid, fumes of PRUSSIC ACID ARE DEVELOPED, THE INHALATION OF THE SMALLEST QUANTITY OF WHICH IS SURE AND SUDDEN DEATH.

Leave everything closed up tight for 40 minutes, then open from the outside and air for at least ten minutes before entering.

HYDROCYANIC ACID.
(For Empty Houses.)

Potassic Cyanide (98 or 99 degrees) one ounce per 100 cubic feet.

Sulphuric Acid (1.83, sp. gr commercial), two fluid ounces per 100 cubic feet

Mix as directed for the last formula

NOTE—I would recommend suburbanites unfamiliar with the handling of these powerful poisons in fumigation not to use them without the help of experts in the work.

SOLUBLE OR MISCIBLE OILS.

Recently the miscible oils have been put upon the market and have now passed the experimental stage and have been found very useful preparations for fighting fruit pests Heretofore the formulae on the market were proprietary and secret, and being too good a thing to be gobbled up by monopolists Mr. L. C Penny has given much attention to devising a practical method for the preparation of miscible oils by means of special soap solutions The soap solution specially recommended by him contains.

Manhadden (fish) Oil, 10 gallons.
Carbolic Acid, eight gallons
Caustic Potash, 15 pounds.

This mixture is heated to about 300 Fahrenheit, after which two gallons each of kerosene and water are added A number of insecticide formulae have been worked out by Mr H Penny, both for winter and summer use The one he considers most efficient for winter use contains 3 2-3 gallons of the above soap solution, 40 gallons paraffine oil, six gallons resin oil, and water sufficient for the desired dilution (See Bulletin 79 of Delaware Agricultural Station)

C O Haughton of Delaware Agricultural College has been experimenting in this direction, and he with other investigators has found it necessary to use kerosene emulsion containing 15 per cent to 20 per cent kerosene, in order to get satisfactory results in destroying scale insects. By means of miscible oils properly prepared scale insects were effectually destroyed when only 10 per cent of heavy oil was present in the spray, and it is believed that a considerably smaller percentage will give satisfactory result. The most effective formula for miscible oil thus far tested calls for nine gallons soap solution, 1¼ gallons water, 40 gallons paraffine oil, and six gallons resin oil. The above quantities are mentioned as being suitable in preparing the oils for orchard use; but as it is only a matter of simple mixing the ingredients a very much smaller quantity would be required for suburbanite's use and may be made by a proportionate reduction in the formula Bulletin 86, Pennsylvania Agricultural Experiment Station, treats also on this subject.

QUASSIA CHIPS.

Quassia Chips, eight pounds
Whale Oil Soap, seven pounds

The quassia chips are boiled in about one gallon of water to each pound of chips for one hour. The soap is added while hot and allowed to dissolve This solution is then diluted with 100 gallons of water. Use with sprayer; or on young trees the tips of the branches affected may be dipped in the liquid This is a very effective wash for the aphis and is much used in Washington, California and Oregon for spraying hop vines, as it is not poisonous.

I may here mention that the lime sulphur wash above mentioned when first introduced was called the Lime, Salt, and Sulphur Wash, as it was made with a proportion of salt, the formula for which was

LIME SALT AND SULPHUR WASH.

Lime, unslaked, 30 pounds.
Sulphur, flowered, 20 pounds
Salt, coarse, 15 pounds
Water, 50 gallons; mix as above

However, when the Agricultural Experiment Stations got to experimenting with it they found that the salt might be dispensed with as superfluous

All these spray formulae are taken from the bulletins, and reports of the United States Department of Agriculture and different Agricultural Experiment Stations and are public property.

In order to make this very important subject more exhaustive I give some extracts from the instructions issued to his customers by one of the largest growers of dwarf trees in Europe.

NOTES on the Prevention and Destruction of Insects, Pests and Diseases affecting Fruit Trees:

Broadly speaking the enemies of plant life may be divided into four classes·

First—The Aphides (Green Fly, etc.)—These have to be destroyed by the direct application of insecticides

Second—Leaf-eating Insects (Caterpillars, Slugs, etc.)—For whose destruction the foliage of the plants must be poisoned.

Three—Fungoid Diseases Mildews, Blister, etc)

Fourth—Mosses, Lichens, Scales and diseases of the bark which can only be effectually dealt with in winter when the trees are dormant.

Before entering into details respecting the above I will enumerate a few remedies which are of easy application Most of those are best applied in a liquid form so that a high class syringe with a fine nozzle is an absolute necessity In large gardens a knapsack spraying pump, such as the Vermorel or Antipest, though apparently costly at first, will save its value in a very short time.

DIRECT INSECTICIDES.

The first among the following is a cheap, effective and easily peated sprayings with formula F (including the Paris Green), first, made insecticide, which I make an extensive use of in my nurseries:

(A) Take half a gallon of paraffine (with a little water added) and two pounds soft soap, and boil these together in an old kettle in the open garden When boiling, carefully skim off the greasy looking surface and then pour the paraffine into a tub containing 25 gallons of rain water Stir thoroughly before using.

(B). Take four ounces of quassia chips and boil them 10 minutes in a gallon of rain water; strain them and add to the liquid four ounces of soft soap, lengthening the whole of 2½ gallons.

(C). Boil two pounds of tobacco leaf, stalks in a gallon of water, strain them and lengthen them to five gallons

LEAF POISONING INSECTICIDES.

(D). The most effective of these is Paris Green Mixtures: This is made by dissolving one ounce of Paris Green and two ounces of fresh lime in 12 gallons of water The lime is needed to destroy the caustic properties of the Paris Green.

(E). Dissolve half pound hellebore powder and half pound alum in tepid water and lengthen to five gallons

ANTI-FUNGOIDES.

(F) The best anti-fungoid is the composition known as Bordeaux Mixture (described above) This and formula D are often used together, in which case four ounces of Paris Green will be required, but no more additional lime.

(G). Another good anti-fungoid spray is made by dissolving five ounces of sulphide of potassium (liver of sulphur) in warm water, and lengthening this to 10 gallons.

(H): If the trees need spraying when the fruit is in a very advanced state it is advisable to use the ammoniacal-copper solution. This is made by dissolving one ounce of copper carbonate in a bottle containing a little under half a pint of ammonia. The bottle will contain sufficient to make 10 gallons of spray. It should be made only as required, as the ammonia exaporates rapidly.

WINTER DRESSING.

(I). Dissolve in separate vessels one pound commercial caustic soda and one pound crude potash, pour the two solutions into a tub containing 10 gallons of water, and add thereto three-fourths of a pound molasses (common treacle) This preparation may be had ready mixed in five-gallon canisters (concentrated) to make 50 gallons of spray.

APPLE AND PEAR DISEASES

Most varieties of fruit have one or more diseases or insects peculiar to them, but on the other hand several of these are common to nearly all kinds of trees They will therefore be referred to under headings of the fruit trees most affected by their ravages.

Green Fly—Apples are frequently and pears occasionally subject to serious attacks from these. They are, however, easily disposed of by spraying with formula.

Woolly Aphis (American Blight) and **Oyster Shell Bark Louse** (Scale) are two of the worst enemies of the bark and young growth. Both are destroyed by painting or spraying the affected part while the trees are dormant, preferably in December and January, with formula I. The former may be kept down in summer by frequent spraying of formula A.

Codlin Moth is probably the most familiar of our apple tree pests, more especially so in the form of "Apple Worm." The moths appear in May and June and lay their eggs at the blossom end of the apple. There they are hatched and after a few days they enter the apple by the crown, making straight for the core. There are three things to be done. First, spray within a week of the fall of the bloom with formula D; second, gather all apples as they fall and destroy them; third, place bands of hay around the stems of the affected trees in July. Remove and burn.

Winter and March Moths—The caterpillars of these, usually known as "loopers," feed on the young foliage of apple trees and occasionally even attack the bloom. It is of the utmost importance that immediately they appear the trees be sprayed with formula D, repeating the dose if necessary a few days later. Prevention, however, is better than cure and a study of their life history shows us that though the male insect has wings the female is practically wingless. As it cannot fly it can only reach the branches of the tree where it may lay eggs by crawling up the stem of the tree. They should therefore be trapped by placing bands of oiled paper smeared with axle grease (or any similar special preparation) around the stems of the trees the second week in October, at which time they are about to commence their upward journey.

Canker—The most frequent and at the same time the most destructive form of canker attacking the apple and pear trees is caused by a minute fungus (Nectria Ditissima) Being unable to pierce the unbroken bark it can only gain admission to the living portion of a branch through a wound. Sometimes these wounds are caused by hail, sometimes they result from punctures of small insects, but in my opinion they are more frequently caused by bursting of unripe wood cells, as explained below.

Having once gained an entrance the fungus spreads rapidly through the bark, which soon shows signs of being eaten away. In the autumn months the presence of the "Nectria" may be recognized by the minute white specks (not to be confused with American blight) which may be seen nestling in crevices of the rugged bark around the edges of the wound. These are the fruits of the fungus which (while in this state) may be destroyed by painting them with a solution of one pound of sulphate of iron, dissolved in a gallon of water. I have given much thought to this subject, fully perused volumes of correspondence in the horticultural press, and have come to the conclusion that in most cases of canker the state of the roots is the secret of the mischief Predisposing causes point to the best way of preventing and restricting its work. It will be noticed that canker seldom appears on trees whose roots are all near the surface, but most frequently on trees which have tap roots whose sap is drawn from the sour subsoil which, owing to its depth from the surface, has not been "sweetened" by the heat of the sun's rays. The cells of the wood produced by such sap cannot be properly ripened and on the appearance of cold the cells give way, causing the mark to split, the "Nectria" at once enters and canker commences its deadly work The remedy, or rather the preventive, is to keep the roots near the surface by frequent mulching of stable manure. Cankered branches should be pared around to the quick and dressed either with clay and cowdung, gas tar or grafting paste.

Pear Midge—This is probably the greatest enemy of the pear. The midge itself is a small knat-like fly, which in April lays its eggs in the opening flower buds of the pear tree without in any way preventing the fruit from setting There are no signs of its presence until a few weeks later, when those attacked commence to swell abnormally and to assume deformed shapes. On examination these will be found to contain a number of small worms. It is imperative that all fruits attacked be gathered and burnt, else early in July they will fall to the ground, turn to chrysalis and give a fresh supply for coming seasons.

Pear Slugs—There are several kinds, but the most plentiful are the grubs of the pear and cherry sawfly. They do a great deal of harm, mostly in May, by eating not only the foliage, but also the tips of the growths of pears and plums. Spray with formula D.

Cracking and Scab occur frequently on trees when the pruning and roots have been neglected A change of soil at the roots will be needed, but the parasite fungi whose present prevent assimilation in the leaf and development in the fruit can only be eradicated by repeated sprayings with formula F (including the Paris Green) First, when the buds begin to swell; second, just before the blooms open; third, when the blooms have fallen, and twice later at intervals of a fortnight.

Mosses, Lichens and Scales—It is a pitiful sight to go through many orchards where apple trees in particular have their bark smothered with moss Such trees cannot possibly grow or bear well. It is essential that their bark be carefully scraped and that in December or January they be thoroughly sprayed with formula I. I cannot too strongly recommend an annual spraying, as it arrests all mossy growth and destroys all scales, besides ridding the trees of insects which hibernate or lay their eggs in crevices or under the edges of the bark.

PLUM AND DAMSON DISEASES.

These are very liable to attacks from GREEN FLY in an aggravated form and will need one or two sprayings of formula A. They are probably more subject to scale than any other fruit tree, and must be sprayed in winter with formula 1, as shown in preceding paragraph.

Red Spider—In both dry seasons the under surfaces of plum leaves are liable to attacks by myriads of these, who suck the sap and choke the leaf pores with their fine webs. A spraying or two of formula A, to which has been added one pound of flowers of sulphur (boiled), will materially assist in eradicating these mites.

Wasps—Plums are probably more than any other outdoor fruit liable to the depredations of wasps. Various wasp poisons are offered, but there is always a certain amount of danger in poisoning the fruit. By far the best method is to trace them to their nest at night. This is easily done by closing up their entrance hole with a soft rag, which has been saturated in a solution made of dissolving 2½ ounces of cyanide of potassium (poison) in 1½ pints of water.

PEACH AND NECTARINE DISEASES.

Blister and Fly Curl—These are erroneously attributed by many to the work of Green Fly, whereas they are in reality caused by a fungoid disease (Exorcus Deformans), which attacks the foliage after a spell of cold winds. The damaged leaves should be picked off and burned and the tree kept clean of flies by spraying of formula A. The fungoid itself is difficult to dislodge and will require at least two sprayings of Bordeaux Mixture (formula F) to cope with its ravages. The first should be applied just before the buds begin to swell and followed by a second and weaker spraying as soon as the flowers have fallen.

CHERRY DISEASES.

Black Fly—Cherries are very liable to attacks from these. As soon as they appear the trees should be sprayed with formula A, or better still, with the tobacco solution C, made slightly stronger. It is often necessary to dip the tips of the branches in the solution in order to destroy the fly.

The falling of cherries at stoning time is usually attributable to the lack of lime in the soil. This may be remedied by freely mixing a quantity of lime rubble with the soil around the roots. A very beneficial autumn dressing for all fruit trees, but especially for stone fruits, consists of 40 ounces Basig Slag and one ounce of Kainit to the square yard as far as the roots extend If the trees are not fairly vigirous this may be followed in early spring by an application of two ounces of Superphosphates and one ounce of Sulphate of Ammonia to the same space.

GOOSEBERRY AND CURRANT DISEASES.

Caterpillars of the gooseberry and currant sawfly may be disposed of by dusting the trees with hellebore powder or spraying them with formula E. Should there be any sign of mildew on the plants the solution of liver of sulphur (formula G) may be sprayed similtaneously with the preceding. Red spider, to which both are liable, should be treated as for plums.

Big Bud—The black currant mite which causes this disease is too small to be seen with the naked eye, but a diseased bud on being examined under the microscope is found to contain myriads of little worm-like insects. As these lay eggs practically all the year round there is no effectual cure. It is advisable to prune off and burn all affected parts and obtain all fresh supplies of black currant bushes from an absolutely untained source.

Gooseberry Mildew—Spray with Bordeaux Mixture as soon as the leaves drop in the fall, again before the buds break in the early spring. When the first leaves have expanded spray with potassium sulphide and repeat at intervals of ten days, if necessary, throughout the summer.

The whole subject of fruit pests and spraying has been exhaustively discussed by the Agricultural Experiment Stations in nearly all the states and their Bulletins will be sent free on demand by citizens of the different states publishing them.

It may be of interest in this connection to give a partial list of some of the Bulletins upon this subject, published by different Agricultural Colleges and Experiment Stations:

Bulletin No. 123, Massachusetts Agricultural Experiment Stations. "Fungicides, Insecticides and Spraying Directions."

Bulletin No. 106, Agricultural Experiment Station of Nebraska. "Does it Pay to Spray Nebraska Apple Orchards?"

Bulletin No. 113, Vermont Agricultural Experiment Station. "Preparation and Use of Sprays"

Bulletin No. 154, Maine Agricultural Experiment Station "Paris Green and Bordeaux Mixture"

Bulletin No. 49, Connecticut Agricultural Experiment station "Petroleum Emulsion for San Jose Scale"

Bulletin No. 23, Montana Agricultural College Experiment Station. "Injurious Fruit Insects Insecticides"

Bulletin No 3, Vol. 4, Pennsylvania Department of Agriculture. "Summer Treatment of Scale Insects."

Bulletin No 296, New York Agricultural Experiment Station, Geneva. "Saving Old Orchards from Scale."

Bulletin No. 95, Arkansas Agricultural Station. "Notes on Spraying"

Circular No. 120, University of Illinois Agricultural Experiment Station. "Spraying Apple Orchards for Insects and Fungi."

In addition to the above the experiment stations in almost every state in the United States have published Bulletins or Circulars more or less elaborate upon this subject, for which readers may apply in writing. However if they will procure the above list and make themselves familiar with their contents they will become fairly well posted in details

VARIETIES OF FRUIT BEST ADAPTED TO THE DWARFING PROCESS.

The vast number of varieties of fruit listed in the nurserymen's catalogues is very confusing to the suburbanite when he requires to make a selection for use. There are, however, some varieties that so much better adapted to the dwarfing process than others, that this appendix may prove helpful. While the commercial orchardist requires only a few varieties, but enough of each to furnish carload lots, and is compelled to conform to the market requirements as to varieties, the suburbanite requires a greater variety and at the same time should plant the best, AND ONLY THE BEST. The following lists may be selected from with the certainty of most satisfactory results. I would say here, however, that these lists by no means exhaust the choicest varieties, but I must draw the line somewhere, and the following will afford ample field to gratify individual fancy:

APPLES.

The following abbreviations are used: (C) for cooking varieties, (D) desert, (C and D) good for both purposes, (F C) show what varieties received the first-class ceritificate of the Royal Horticultural Society of England, which is the highest award given by that society and is a guarantee of the highest quality. (*) signifies extra quality, (**) double extra, and (***) of superlative excellence.

SELECTED DOZENS FOR SPECIAL QUALITIES.

Large Size.

ALFRISTON (C).
BISMARCK (C FC *).
BRAMLEY'S SEEDLING (C FC *).
ECKLINVILLE SEEDLING (C).
EMPEROR ALEXANDER (C).
GLORIA MUNDI (C).
LORD SUFFIELD (C **).
MERE DE MENAGE (C).
MONSTREUSE INCOMPARABLE.
PEASGOOD'S NON-SUCH (C).
POTT'S SEEDLING (C).
WARNER'S KING (C *).

Bright Color.

BISMARCK.
CELINEE (C *).
COX'S POMONA (C).
DEVONSHIRE QUARENDEN (D *).
EMPEROR ALEXANDER.
GASCOIGNE'S SCARLET (C FC).
HOLLANDBURY'S ADMIRABLE (C).
LADY HENNEKER (C D FC).
MERE DE MENAGE.
RED ASTRACHAN (C D).
THE QUEEN (C D FC).
WORCESTER PEARMAIN (C D FC).

Fine Flavor.

ALLINGTON PIPPIN (D FC).
BLENHEIM ORANGE (C D *).
CORNISH GILLIFLOWER (D).
COX'S GOLDEN PIPPIN (D).
DUKE OF DEVONSHIRE (D *).
GOLDEN PIPPIN (D *).
IRISH PEACH (D).
KING OF PIPPINS (D *).
MARGIL (D **).
MR. GLADSTONE (D *).
RIBSTON PIPPIN (D).
ROYAL RUSSET (D).

Heavy Crops.

ALFRISTON.
BISMARCK.
CELINEE.
DEVONSHIRE QUARENDEN.
ECKLINVILLE SEEDLING.
HAWTHORNDEN (C).
KESWICK CODLIN (C).
LANE'S PRINCE ALBERT (C FC *).
LORD SUFFIELD.
POTT'S SEEDLING.
STIRLING CASTLE (C).
WORCESTER PEARMAIN.

Some of the above excel in more than one quality and consequently are more desirable. Detailed description of the above will be found further on. While the above may prove a sufficient list for the majority of suburbanites to select from, I will add a descriptive list of 50 of the best quality of apples in order to furnish a wider range for selection.

Descriptive List of Fifty Best Apples.
(For Suburbanite's Use.)

DESSERT.

ALLINGTON PIPPIN (F C)—A richly flavored apple, result of cross between King of Pippins and Cox's Orange Pippin. Yellow streaked with red on sunny side, good bearer, free grower, November to February

BELLE FLOWER (*)—Large and excellent; skin smooth, yellow, tender, juicy, crisp. November to January

BLENHEIMS ORANGE—Flesh yellowish, crisp, juicy, good both for table and kitchen use. November to February.

CHARLES ROSS (F C)—A seedling from Cox's Orange Pippin, large and handsome, solid, heavy, and high flavored. November to December

CLAYGATE PEARMAIN—Medium size, richly flavored, highly aromatic, similar to Ribston Pippin. January to May

COX'S ORANGE PIPPIN (***)—Medium size. There is no better apple grown. October to February

CORONATION—Medium size, resembling Cox's Orange Pippin, suffused with red and streaked on sunny side. September to October

COURT PENDU PLAT (*)—Medium, handsomely shaped. A valuable desert apple of first quality. In use December to May

DETROIT RED—Above medium, entirely covered with uniform darkest red, flesh suffused with bright red, of very fine flavor. November to January

DEVONSHIRE QUARENDEN—Medium, skin almost entirely dark purplish red, crisp, juicy and rich, the best early high colored apple. August

DUKE OF DEVONSHIRE—Medium, rich, crisp, juicy. February to May.

GRAVENSTEIN (*)—Large, popular, high quality, good also for kitchen. September to January

GOLDEN PIPPIN—Well known for excellence, small size. November to February

IRISH PEACH—Early and high perfumed, juicy and well flavored, medium size. August.

KING OF PIPPINS—Medium to large, very handsome, crisp and juicy. October to January

MARGIL—Small, richly flavored, one of the finest desert apples. November to March.

MANNINGTON'S PEARMAIN—Medium, juicy, sweet, flavor rich, should be allowed to hang late on the tree. November to March.

McINTOSH RED (*)—Medium, hardy, nearly covered with dark red; flesh white, fine, juicy and refreshing. November

MR. GLADSTONE (F C)—The earliest desert apple; mottled red, with yellow streaks, carries bloom like a plum, very prolific. July

RIBSTON PIPPIN—Medium, a favorite English apple, flesh yellow, firm, with rich aromatic flavor, very prolific. October to January

RED JUNEATING—Medium size, early and excellent, very popular. July.

SPITZENBERG ESOPUS—Medium to large, deep red, flesh yellow, highly flavored, sub-acid. November to March.

STURMER'S PIPPIN—Medium size, firm, brisk, richly flavored, a valuable late keeping apple, coming into use when other late varieties are over February to June

THE HOUBLON—A grand desert apple, from a pip of the same fruit from which the Charles Ross was raised, a deeper and longer keeping Cox's Orange Pippin, remarkably firm fleshed and attractive variety. October to March

THE QUEEN (F C)—Handsome, a great bearer, flesh white tender and excellent; enormously prolific September

WINTER BANANA (***)—This is a remarkably handsome apple, large, with strong banana flavor The highest priced apple on the market October to July

CULINARY.

ALFRISTON—A good bearer, and one of the largest apples in cultivation November to April.

BEAUTY OF BATH (F C)—A beautiful striped early apple, good grower and abundant bearer August

BISMARCK (F C)—One of the handsomest apples in cultivation and a profuse bearer November to February

BRAMLEY'S SEEDLING (F C)—Skin striped with scarlet; very large, flesh firm, acid and juicy, a valuable late apple December to March

CELENEE—A handsome large red apple, one of the most commendable of all apples October and November

CANADA REINETTE—Large yellow, firm, well flavored, good for desert or kitchen, an abundant bearer November to January

COX'S POMONA—Large, very handsome, one of the best and most prolific of apples September to October

DUTCH MIGNONNE—Large, round, smooth and handsome, good for table or kitchen December to April

DUMELOW'S SEEDLING (Wellington)—Large and excellent, always retains its acid November to March

ECKLINVILIE SEEDLING—Large, roundish, a handsome and excellent apple and great bearer October to January

EMPEROR ALEXANDER—Fruit very large, roundish ovate, flesh tender; one of the most beautiful apples September to November

GASCOIGNES SEEDLING (F C)—A distinct richly colored and exceedingly handsome apple, free bearer, its bright color makes it a favorite for the table and kitchen. November to March

GOLDEN NOBLE—Very large, roundish, skin smooth, clear bright yellow, flesh yellowish, firm and juicy October to March

HAWTHORNDEN—Large, roundish yellow, flesh white, crisp, tender and very highly flavored October to January.

KESWICK CODLIN—Medium, conical, angular, light yellow, a great bearer, seldom misses a crop. August and September

KING OF TOMPKINS COUNTY—Very large and handsome, flavor rich, tender and good, equally adapted to table or kitchen November to January

LANE'S PRINCE ALBERT—An extremely handsome and late variety, a good bearer, very recommendable October to January

LORD SUFFIELD—Very large, nearly white; a most abundant bearer. August and September

NEWTON WONDER—Large, solid, keeps late, very prolific, one of the best of recent introductions November to May

PEASGOOD'S NON-SUCH (F C)—One of the handsomest apples in cultivation; flesh yellow, tender, juicy, with an agreeable acid flavor October to January

POTTS' SEEDLING—A handsome and prolific yellow apple; a very desirable variety September and October

STERLING CASTLE—An excellent apple and great bearer, skin green, turning to pale yellow, flesh white, tender and juicy. October and November

WEALTHY—Very hardy, prolific, medium to large size, red streaked, good packer on account of uniform shape, good for table or kitchen   September to December

WARNER'S KING—Very large, deep yellow strewed with russet; flesh white, tender, crisp and juicy, with fine brisk sub-acid flavor; first-rate culinary apple, prolific. October to January.

WORCESTER PEARMAIN (F C)—Medium size, skin completely covered with red, flesh tender, juicy and well flavored, early and of good quality. August and September

While the above is a good list of choice variety of apples that will dwarf well, and give satisfaction, yet there are lots more that might be added. In one of the nurserymen's catalogues now before me 172 varieties of apples are listed, and others give still longer lists. It may be that many of my readers are unfamiliar with the names of some of the varieties, they being of European origin. This arises from the fact that more attention has been paid to dwarfing there, and I avail myself of the European's well practiced experience. At the same time I include a number of choice American varieties.

PEARS.

To France and the Island of Jersey we owe some of our best pears, and it will be noticed that in the following list I have drawn largely on both these countries, as the dwarfing system has been highly developed there   (*) means extra, (**) means a super-excellent. (C) for culinary. (D) for dessert, (DC) good for both purposes, (DG) double grafted

BARTLETT (Bon Cretien)—A well known pear of the highest excellence. August and September

BELLE DE JERSEY (BELLE ANGEVENE)) (C)—This is the largest of all pears, sometimes weighing up to three pounds, is very ornamental and is used in France more to ornament the dinner table than for eating It frequently sells in the Palais Royal, Paris, for 30 francs each. November to May

BERG-AMOTTE ESPEREN (D)—A delicious late pear, medium size, flesh yellowish, melting and juicy February to April

BEURRE D' ANJOU (D)—Large, an excellent melting pear; one of the best early winter pears   December

BEURRE D' AMANLIS (* D)—This is one of our best September pears; very large, tender, juicy, melting, with richly perfumed flavor; an excellent wall pear   September

BEURRE D' AREMBERG (D)—Very juicy, sweet and delicious with pleasant aroma   September   (Known as GLOU MORCEAU in England)

BURRE BEURRE DIEL (D)—Very large, aften weighing 16 to 20 ounces; melting and excellent; a well known pear. November and December

BEURRE GIFFORD (D)—Medium size, melting, very juicy, one of the best early pears, a good pear   July

BEURRE HARDY (D)—Large, an excellent melting pear, remarkable for its beauty and vigor of growth on the quince, is very good on a wall   October.

BEURRE RANCE (D G)—Often very large; a most excellent late melting pear, requires double grafting; forms a better bush than pyramid. March and April,

BEURRE SUPERFINE (* D)—A most delicious desert pear; well known as one of our best dessert pears. September and October

BEURRE CAPLAUMONT (D C)—Medium size, juicy and agreeable, a most abundant bearer October

BEURRE LUCRATIVE (D)—A fine medium, melting pear, yellow, delicious, good grower and productive September

CATALLAC (* C)—An immense bearer, best for stewing, owing to large size, is better as bush or espalier December to March

CLAIRGEAU (D G)—Large and very handsome; double grafted, it makes a handsome pyramid December

CLAPS FAVORITE (* D)—An American pear of high excellence, large, handsome and exceedingly good, and is valuable for either wall or orchard August.

CONFERENCE (* D FC)—Fruit large; flesh salmon colored, melting, juicy and rich, tree robust and hardy. Early November

CITRON DE CARMES (D DG)—Below medium size, when double grafted it bears very fine fruit, is very popular in France, where few fruit gardens are without it July

DURONDEAU (DE TONGERS)—Very large and handsome, melting, rich and delicious, a good wall or espalier pear October and November.

DUCHESSE D' ANGOULEME (D C)—A very large and noble looking fruit, best suited for exhibition, though when grafted on quince the fruit becomes melting and rich. October and November

DR JULES GUYOT (* D)**—A great improvement on the Bartlett, being earlier, is often of the highest quality. August and September

DOYENNE BOUSSOCH—A very large and handsome pear, which succeeds and bears profusely on the quince, good on wall or trellis October and November

DOYENE DE COMICE (D C)**—A most delicious pear, of largest size, very handsome, melting and juicy. The fruit on a wall or espalier is superb in quality and appearance In the orchard house, in a pot, the fruit will ripen on the tree into November and may then be gathered and eaten November.

FORELLE (TROUT PEAR) (D G)—Medium, a very handsome speckled pear; succeeds well double grafted; the color of skin is very attractive November.

GENERAL TODLEBEN—Very large, melting and juicy; good as a wall or espalier pear, great bearer November.

JERSEY GRATIOLI—A delicious pear and great bearer. September and October

JERSEY CHAUMONTEL (D)**—The best and finest of all our pears, bearing a crop when all other fail, the flavor of this remarkable pear is unequalled, often weighs one and one-half pounds

JARGONELLE (D G)—A large, well known early variety, particularly adapted for an early wall July.

JOSEPHINE DE MALINES (*)—Medium size, a delicious hardy, melting pear, with rich aromatic flavor, succeeds well on quince, bears well as bush or espalier, is very prolific January, February, March.

GANSELS BERGAMOT (D G)—Large and very handsome, perfumed, melting and excellent; slow in coming into bearing, unless double grafted October and November

LOUISE BON DE JERSEY (D)—A general favorite, beautiful and good. October

LAWRENCE D *)—A very late, long keeping, medium size pear of high quality. December to March

LE LECTIER (* D)—A large French staple pear of high quality, tree vigorous and constant bearer January to March

MARGARET MARILLAT (D)—Large, handsome, with distinct flavor September

MARECHAL DE LA COUR (D)**—Large, hardy, melting, very fine on wall or trellis. October and November

MARIE LOUISE (* D)—Large, well known pear of highest excellence; double grafted on wall or espalier it attains large size, and is invariably of good flavor October and November

MADAME TREYVE (* D)—Large, early, melting and very rich, hardy. September

PITMASTON DUCHESSE, (* D C)—Very large, of good quality, good at all points—in orchard house, on a wall, as an espalier, or as an orchard standard October and November.
PASSE CRASANNE (D)—An excellent late pear, one of the finest, requires a good soil and high culture, and warm situation to develop its best qualities January to March.
PRINCE NAPOLEON (D)—Large, a seedling from Passe Crasanne, melting, with fine aroma January to March
SECKLE (** D G)—Small, an American pear, allowed to be "the standard of excellence in pears" October and November
SOUVENIR DU CONGRESS (** DG D C)—Very large, weighing one to two pounds, on quince requires double grafting, excellent on a wall September.
SHELDON (* D)—Large, globular, russet, flavor resembling the Seckle October to December
SWAN'S ORANGE (ONODAGA) (D C)—Large yellow pears of good quality. November to December
WINTER ORANGE (C FC)—Medium to large, rich russet brown January
WINTER NELLIS (* D)—Below medium, melting, juicy, delicious flavor; productive December to January.

PEACHES.
(Best Twelve)

ALEXANDER (** Semi-Cling)—A handsome, richly colored peach, very popular among fruit growers, hardy, ripening out of doors about the middle of June (FC?). July
CRAWFORD EARLY (Free)—A magnificent, large, yellow peach of a good quality; its size, beauty and productiveness makes it one of the most popular varieties, hardy Midseason
CRAWFORD, LATE (*? FREE)—Fruit large, yellow, vigorous, productive, one of the finest late sorts, hardy End of September
CARMAN (*? FREE)—Large, resembling Elberta pale yellow, prolific, hardy. Midseason
CHAMPION (**? Free)—Of recent introduction, large, of high quality; hardy. Ripens after Early Crawford
ELBERTA (**? FREE)—Very large, golden yellow, hardy, prolific. A general favorite Late September.
GROSSE MIGNONNE (* Free)—Very large and very good, midseason peach September
HALE'S EARLY (*** Free)—A magnificent early peach, highly colored and richly flavored, by far the most popular of the early peaches Ripens in July
NIAGARA (**? Free)—A large, beautiful and high quality peach; hardy End of August
NOBLESSE (** Free)—One of the best peaches; very large, one of the best either for forcing or open wall September
PRINCESS OF WALES (** Free)—Very largest and best of peaches known Middle to end of September
ROYAL GEORGE (** Free)—A great favorite, large, a most reliable mid-season peach August and September

NECTARINES.

ADVANCE (*)—One of the earliest Nectarines August.
EARLY RIVERS (* FC)—The earliest of the nectarines, a grand acquisition Beginning of August.
LORD NAPIER (* Free)—Fruit very large and handsome August
PITMASTON ORANGE (Free)—Large, rich and sweet September.

PINEAPPLE (Free)—Large and very rich, ripens a week after the Pitmaston September

STANWICK ELRUGE (* Free)—Large and richly flavored; flesh white and sugary September

APRICOTS.

BLENHEIM (*)—Medium, juicy and good; tree hardy and not liable to gum September

MOOR PARK (* Free)—Large, early, juicy, rich and excellent, tree hardy End August.

ROYAL (* Free)—A standard variety, of great hardiness and all around good qualities. July and August.

EUROPEAN GRAPES.
(Hardy for Outdoors.)

BLACK HAMBURG—The most popular European grape in cultivation; very large, juicy, vinous and rich; most popular variety

GROS COLMAN—Best late grape, of noble appearance, easily cultivated; flavor improves by late keeping

LADY DOWNE'S SEEDLING—Valuable late keeping grapes of fine flavor

SWEET WATER—The best of the hardy grapes, succeeds well in the open.

AMERICAN GRAPES.

NIAGARA—Hardy, white
CONCORD—Hardy, black
MOORE'S EARLY—Hardy, black

BRIGHTON.
DELAWARE.
AGAWAM.

CHERRIES.

BLACK BIGARREAU—An excellent large black cherry, very recommendable July and early August.

BLACK TARTARIAN—Very large, sweet and good, a good bearer. Ripe July and August

BLACK REPUBLIC—Large, medium season, a good shipper August.

BING—A strong grower, fruit very large, very hardy and productive, fine market variety. End July

DIKEMAN—A large dark cherry; hangs long on tree. August

EARLY RICHMOND—Red, acid, juicy, good for cooking June

ELTON—Very large, light red, and excellent July

ENGLISH MORELLO—Productive, and late, a culinary cherry of good quality; when grown on a north wall of building may be kept in good condition until September and October

GOVERNOR WOOD—Good, early, light cherry; tender, juicy, sweet and delicious. End of June

LAMBERT—Very large, flesh firm, flavor unsurpassed; excellent shipper; ripens two weeks later than Royal Anne End of August

MAY DUKE—Best of the early cherries, well known old variety June.

NOBLE (F C)—Large, very dark red, flesh firm and very late, a new sort. September

OLIVET—Large, globular, very shining; deep red, flesh tender, rich, vinous, with sweet sub-acid flavor; one of the most profitable and latest cherries. Sept

ROYAL ANNE (NAPOLEON BIGARREAU)—A very fine cherry of largest size; very productive August

STRANG LOGIE—A magnificent early dark red cherry; rich flavor and an extraordinary bearer. June and July

WHITEHEART—A beautiful cherry; pale yellow, marbled with red, a great bearer, certainly one of the very best cherries Middle of July.

WINDSOR—A valuable late market cherry; fruit large, liver colored, flesh remarkably firm and of fine quality; very prolific End of August.

PLUMS.

REINE CLAUDE DE BAVAY—Round, greenish yellow, very large, rich and delicious October

BELGIAN RED—A delicious dessert plum End of August

COLUMBIA—Very large, round, purple, rich and sugary, parts freely from the stone August

CZAR—The earliest of the fairly large blue plums; the best of its season. End of July.

COE'S GOLDEN DROP (SILVER PRUNE)—One of the richest yellow plums, very large. End of September.

ITALIAN PRUNE (HELLEMBERG; Large German Prune)—Large, juicy and delicious, freestone, excellent for drying September

JEFFERSON—A fine, large, oval, yellow plum, very rich, juicy, freestone August

KIRKE'S—A delicious dessert plum fruit, large, purple, with blue bloom, that does not easily rub off, firm, juicy and very richly flavored. September

MONARCH (F C)—Fruit very large, dark purplish blue, freestone, of excellent quality, skin does not crack with heavy rain; grown on a wall the fruit grows very large Late September

PEACH PLUM—Very large, brownish red, coarse grained but juicy and pleasant. July.

POND'S SEEDLING (HUNGARIAN PRUNE)—An enormous bright red culinary plum, decidedly one of the best for cooking purposes; very productive and a good shipper. September

SHROPSHIRE DAMSON—Small, oval, very prolific; culinary September.

VICTORIA—The most prolific of all plums; fruit large, red, juicy, sweet and pleasantly flavored September.

WASHINGTON—Large, yellow, marked with crimson dots; rich and sugary September.

FIGS.

BROWN TURKEY—The hardiest variety, brownish purple; large, rich and excellent, bears most abundantly in pots and on walls and forces well.

WHITE MARSEILLES—Large, greenish white, of most luscious sweetness; bears abundantly and forces well

CURRANTS.

Black.

BLACK NAPIES—Very large and good

LEE'S BLACK—Large, very productive, very sweet

VICTORIA—Black, large and sweet, with long bunches.

Red.

FAY'S PROLIFIC—The largest red currant, bunches short, and very good.

LA VERSAILLAISE—Very large, and good; an abundant bearer

RED DUTCH—Bunches short, a sweet, rich and good currant.

White.

WHITE DUTCH—A well known good sort.

NOTE—To produce very large red and white currants the bush should be cut in closely, i. e. the young shoots should annually be shortened to two inches. Currants make very handsome pyramids and bear profusely.

GOOSEBERRIES.
Rough Reds.

CHAMPAIGNE—Very rich flavor
COMPANION—Extra fine
CROWN BOB—A sure cropper and good quality.
IRONMONGER—Small, high flavor.
LANCASHIRE LAD—Great bearer.
RIFLEMAN—Immense, late.
VICTORIA—New, highly prized
WARRINGTON—Good flavor, late.
WINDHAM'S INDUSTRY—Large size

Smooth Reds.

CONQUERING HERO—Heavy cropper
MAJOR HIBBERT—Large.
SLAUGHTERMAN—Extra fine
MAYDUKE—Very early and desirable.

Smooth Yellows.

LEADER—Very fine.
LEVELLER—Extra large, good flavor

White.

WHITE SMITH—Earliest white.
ANTAGONIST—The largest white

Green.

BERRY'S EARLY KENT—Early, delicious
DRILL—Extra fine.
HARRABY EARLY GREEN—The earliest.
KEEPSAKE—Large, early, delicious
LANCER (Howard's)—Large and great bearer
LANGLY'S GAGE—Highly flavored; grand for dessert.
SHAKESPEARE—Very large
SIR GEORGE BROWN—Large and well flavored
STOCKWELL—Large and good.
TELEGRAPH—Extra fine; slow grower.

Rough Yellows.

BROOM GIRL—Large and good flavor
CATHERINA—Long and fine
GOLDEN DROP—Very early, fine flavor
GUNNES—Handsome, richly flavored
LANGLY BEAUTY—Large and good; highly flavored

DR. A. W. THORNTON'S
Suburbanite's Dwarf Fruit Tree Nursery
FERNDALE
Whatcom County, Washington

This nursery is being established for the propagation of Dwarf Fruit Trees and will be confined to that class of fruit and will be conducted on the co-operative plan and strictly under the Golden Rule, a large proportion of the profits being set aside and divided annually among the employees.

The operations of the nursery will be confined to the production of only the highest quality of fruits and such as have been found to respond best to the dwarfing process No inferior stock will be allowed to grow in the nursery or sold therefrom.

The classes of fruit grown will consist of

POME FRUITS—(Apples, Pears and Quinces)

STONE FRUITS—(Peaches, Nectarines, Apricots, Cherries and Plums)

SMALL FRUITS—Currants, Gooseberries, Raspberries, Blackberries, etc.)

TRAINED TREES—Special attention will be given to this line.

This nursery will be kept up to date in every department and the interests of suburbanites will be a fundamental rule

For further particulars and price lists, apply to as above.

Suburbanite's Special Collection of Dwarf Fruit Trees

I will furnish these special collections at the prices named. Delivered free by mail for cash with order. The trees will be of the best varieties, assorted of the class known as "Maidens," or one year from the bud, and pruned back ready for planting

If desired, trees of a bearing age (2 years) and furnished with fruit buds that will bloom and bear fruit the first season after planting will be furnished F. O. B. at Ferndale, Whatcom County, Washington (as the roots would be too well developed to send safely by mail), at 50 per cent above the price of "Maidens."

Collection A—Apples.

Five assorted apples (early and late, culinary and dessert); my own selection of varieties; all maidens; free by mail..........$4.50

Collection B—Apples.

Five similar varieties, 2 years old (bearing age), F. O. B. at Ferndale, Whatcom County, Washington....................$6.75

Collection C—Pears.

Five pears, assorted varieties, Maidens; my selection; free by mail $4.50

Collection D—Pears.

Five pears, similar varieties, 2 years old (bearing age), with fruit buds; F. O. B. at Ferndale............................$6.75

Collection E—Stone Fruit.

Five assorted stone fruit (peaches, nectarines, apricots or cherries); my selection; "Maidens;" free by mail for...............$9.00

Collection F—Stone Fruit.

Five assorted stone fruit, as above, 2 years old (bearing age); F. O. B. at Ferndale, Whatcom County, Washington, for...... $12 50

SMALL FRUIT.
Collection G—Gooseberries.

One dozen assorted Maidens, gooseberries, red, white, yellow, green; prize varieties, my selection, free by mail (these will be adapted to training as upright cordons or U form cordons).....$3.00

Collection H—Gooseberries.

One dozen assorted gooseberries, 2 years old, bearing age; F. O. B. at Ferndale $4 50

Collection I—Currants.

One dozen assorted currants (white, red or black), Maidens; free by mail $2 50

Collection J—Currants.

Six fancy trained assorted currants (bushes—standards), cordons; fruiting age... $3 00

DR. A. W. THORNTON
Suburbanite's Dwarf Fruit Tree Nursery
Ferndale, Whatcom County
Washington

www.ingramcontent.com/pod-product-compliance
Lightning Source LLC
Chambersburg PA
CBHW050803271025
34568CB00011BA/922